TOTAL PRODUCTIVE MAINTENANCE

TERRY WIREMAN

INDUSTRIAL PRESS

Library of Congress Cataloging-in-Publication Data
Wireman, Terry.
   Total productive maintenance/ Terry Wireman.--2nd ed.
      p. cm.
   ISBN 0-8311-3172-1
   1. Plant maintenance--United States--Management. 2. Industrial equipment--United States--Maintenance and repair--Management. 3. Total productive maintenance. I. Title.

TS192.W575 2004
658.2'02--dc22   2003064093

Industrial Press, Inc.
200 Madison Avenue
New York, NY 10016-4078

First Edition, May 2004

Sponsoring Editor: John Carleo
Interior Text and Cover Design: Janet Romano
Developmental Editor: Robert Weinstein

Copyright © 2004 by Industrial Press Inc., New York. Printed in the United States of America. All right reserved. This book, or any parts thereof, may not be reproduced, stored in a retrieval system, or transmitted in any form without the permission of the publisher.

Printed in the United States of America

10   9   8   7   6   5   4

TABLE OF CONTENTS

Part 1

Chapter **1** The History and Impact of Total Productive Maintenance   1

Chapter **2** Defining Equipment Losses   30

Chapter **3** Overall Equipment Effectiveness   47

Chapter **4** Activities Before TPM   59

Part 2

Chapter **5** Developing the TPM Implementation Plan   81

Chapter **6** Preventive Maintenance   90

Chapter **7** Maintenance Inventory Controls   108

Chapter **8** Improving Maintenance Efficiency and Effectiveness   122

Chapter **9** Maintenance Information Systems   129

Chapter **10** Capacity Assurance Technicians   145

Chapter **11** Total Economic Maintenance   154

Chapter **12** Team-Based Maintenance   168

Chapter **13** Performance Indicators for TPM   176

Chapter **14** The Future of TPM   181

Chapter **15** Maintaining the TPM Vision   184

Index   195

Preface

Total Productive Maintenance, or TPM for short, is an advanced manufacturing technique that focuses on maximizing the overall equipment effectiveness of any asset utilized in the production of goods and services. While the basic components of TPM have been in existence for decades, few companies are able to assemble the components into an overall strategy. Many companies will partially implement some of the components, but never realize the full benefits that can be achieved through TPM.

One of the primary issues companies face while attempting to achieve TPM is the organization's perception of TPM. Many companies will fail to achieve results because of the paradigm around the word "Maintenance." For example, in some organizations maintenance is described with words such as:

> Overhead
> Necessary Evil
> Expense
> Prima Donnas
> Fire Fighters
> Janitors

These organizations have a negative paradigm about the maintenance function. In contrast, some organizations describe maintenance with words such as:

> Preventive
> Predictive
> Reliability
> Technicians

These organizations have a positive paradigm about maintenance and will be able to achieve TPM results in a more rapid timeframe.

Organizations with a negative paradigm about maintenance may take three to five years, and perhaps even longer, to fully achieve the benefits of TPM. Some organizations have even found that the maintenance function was viewed so negatively they had to rename the function before they could make progress. For example, some companies use the acronyms:

> TPR: Total Process Reliability
> TPM: Total Productive Manufacturing
> CAM: Capacity Assurance Management

Ultimately, it does not matter what the process is called. What really matters are the results that are achieved. The purpose of this text is to outline a methodology that can be used to implement TPM successfully. Whether a company uses another acronym to describe TPM does not matter. When all is finished, the results are what matter.

It is hoped that by applying many of the suggestions contained in this text, companies can not only implement TPM, but also achieve the financial benefits that are available.

PART 1

PREPARING FOR TOTAL PRODUCTIVE MAINTENANCE

In this portion of the book, we will examine the activities necessary for a company to prepare for *Total Productive Maintenance* (TPM). These activities include a clear understanding of what TPM really is as well as what it is not. We will examine the current state of maintenance management in the United States, while illustrating why improvement is vitally important to corporations. We will also highlight pre-implementation activities that must be conducted before any TPM program can be successful.

CHAPTER 1

The History and Impact of Total Productive Maintenance

In order to properly understand the history and impact of total productive maintenance, it is necessary to establish a definition. *Total Productive Maintenance* (TPM) is maintenance activities that are productive and implemented by all employees. TPM involves everyone in the organization from operators to senior management in equipment improvement. It encompasses all departments including:

- Maintenance
- Operations
- Facilities
- Design Engineering
- Project Engineering
- Construction Engineering
- Inventory and Stores
- Purchasing
- Accounting and Finance
- Plant and Site Management

Goals of TPM

TPM has the following five goals (some texts call these pillars):
1. Improving equipment effectiveness
2. Improving maintenance efficiency and effectiveness

3. Early equipment management and maintenance prevention
4. Training to improve the skills of all people involved
5. Involving operators (occupants) in routine maintenance

Improving Equipment Effectiveness

This goal, which insures that the equipment performs to design specifications, is the true focus of TPM. All remaining goals for TPM are valueless unless they support improving equipment effectiveness. The focus must be that nowhere in the world can another company have the same asset and make it produce more than your company can produce. If it does, then it is better at managing its assets than your company and will always be the lower cost producer or provider.

The equipment must operate at its design speed, produce at the design rate, and produce a quality product at these speeds and rates. A major problem occurs because many companies do not know the design speed or rate of production for their equipment. In the absence of knowing the design criteria, management will set arbitrary production quotas. A second major problem develops over time when small problems cause operators to change the rate at which they run equipment. As these problems continue to build, the equipment output may only be half of that for which it was designed. This inefficiency then leads to the investment of additional capital in equipment, trying to meet the required production output.

Improve Maintenance Efficiency and Effectiveness

This goal focuses on insuring that maintenance activities that are carried out on the equipment are performed in a way that is cost effective. Studies have shown that nearly one-third of all maintenance activities are wasted. Therefore, this goal of TPM is important to lowering the cost of maintenance. It is important for all to understand that basic maintenance planning and scheduling are crucial to achieving low-cost maintenance. The goal is to insure lean maintenance, with no waste in the maintenance process.

A secondary goal is to ensure that the maintenance activities are carried out in such a way that they have minimal impact on the up time or unavailability of the equipment. Planning, scheduling, and backlog control are again all important if unnecessary maintenance downtime is to be avoided. At this stage, maintenance and operations must have excellent communication in order to avoid downtime due to misunderstandings.

Developing an accurate database for each piece of equipment's maintenance history is also the responsibility of the maintenance department. This history will allow the maintenance department to provide accurate data for decisions related to the plant or facility equipment. For example, the maintenance department can provide input to equipment design and purchase decisions, assuring that equipment standardization is considered. This aspect alone can contribute significant financial savings to the company. Standard-ization reduces inventory levels, training requirements, and start-up times. Accurate equipment histories also helps stores and purchasing not only reduce downtime, but also avoid carrying too much inventory.

Early Equipment Management and Maintenance Prevention

The purpose of this goal is to reduce the amount of maintenance required by the equipment. The analogy that can be used here is the difference in the maintenance requirements for a car built in 1970 compared to a car built in 2000. The 1970 car was tuned up every 30-40,000 miles. The 2000 car is guaranteed for the first 100,000 miles. This change was not brought about by accident. The design engineers carefully studied the maintenance and engineering data, allowing changes to be made in the automobile that reduce the amount of maintenance. The same can be true of equipment in a plant or facility.

Unfortunately, most companies do not keep the data necessary to make these changes, either internally or through the equipment vendor. As a result, unnecessary maintenance is performed on the equipment, raising the overall maintenance cost.

Training to Improve the Skills of All People Involved

Employees must have the skills and knowledge necessary to contribute in a TPM environment. This requirement involves not only the maintenance department personnel, but also the operations personnel. Providing the proper level of training insures that the overall equipment effectiveness is not negatively impacted by any employee who did not have the knowledge or skill necessary to perform job duties.

Once employees have the appropriate skills and knowledge,

their input on equipment improvement needs to be solicited by senior management. In most companies, this step only takes the form of a suggestion program. However, it needs to go well beyond that; it should also include a management with an open doors policy. Such a policy indicates that managers from the front line to the top are open and available to listen to and give consideration to em-ployee suggestions. A step further is the response that should be given to each discussion. It is no longer sufficient to say "That won't work" or "We are not considering that now." In order to keep communication flowing freely, reasons must be given. Therefore, managers must develop and utilize good communication and management skills. Otherwise, employee input will be destroyed and the ability to capitalize on the greatest savings generator in the company will be lost.

Involving Operators (Occupants) in Routine Maintenance

This goal finds maintenance tasks related to the equipment that the operators can take ownership of and perform. These tasks may amount to anywhere from 10-40% of the routine maintenance tasks performed on the equipment. Maintenance resources that were formerly engaged in these activities can then be redeployed in more advanced maintenance activities such as predictive maintenance or reliability focused maintenance activities. It must be noted: the focus for the operations involvement is *not* to downsize the maintenance organization. Instead, the focus is to free up maintenance resources for the more technical aspect of TPM.

Cost-Benefit of These Goals

The questions now raised are: Are these goals all worth it? What are the benefits that have been achieved? These questions are answered positively and quickly because results are as follows:

Productivity
100-200% increases
50-100% increase in rates of operation
500% decrease in breakdowns

Quality
100% decrease in defects
50% decrease in client claims

Costs
  50% decrease in labor costs
  30% decrease in maintenance costs
  30% decrease in energy costs
Inventory
  50% reduction on inventory levels
  100% increase in inventory turns
Safety
  Elimination of environmental and safety violations
Morale
  200% increase in suggestions
  Increased participation of employees in small group meetings

With all of these benefits, it is important for all companies to recognize the importance and value that productive maintenance can bring to the company. Any company trying to achieve World Class status through other programs such as Computer Integrated Manufacturing (CIM), Just in Time (JIT), Total Quality Control (TQC), Total Employee Involvement (TEI), or Lean Manufacturing, will soon find that these programs will not work without total reliability of the company's assets, which is the primary responsibility of the maintenance organization. In particular, Just in Time, Total Quality Control, and Total Productive Maintenance are all essential. Without full utilization of these three programs, the goal of being globally competitive will never be reached.

History of TPM

From where did TPM evolve? What spurred its development? TPM originated in Japan and was an equipment management strategy designed to support the Total Quality Management strategy. The Japanese realized that companies cannot produce a consistent quality product with poorly-maintained equipment.

TPM thus began in the 1950s and focused primarily on the preventive maintenance. As new equipment was installed, the focus was on implementing the preventive maintenance recommendations by the equipment manufacturer. A high value was placed on equipment that operated at designed specifications with no breakdowns. During these same years, a research group was formed which later became the Japanese Institute of Plant Management (JIPM).

During the 1960s, TPM focused on productive maintenance, recognizing the importance of reliability, maintenance, and economic efficiency in plant design. This focus took much of the data collected about equipment during the 1950s and fed it back into the design, procurement, and construction phases of equipment management. By the end of the 1960s, JIPM had established and awarded a PM prize to companies that excelled in maintenance activities.

Then in the 1970s, TPM evolved to a strategy focused on achieving PM efficiency through a comprehensive system based on respect for individuals and total employee participation. It was at this time that "Total" was added to productive maintenance. By the mid-1970s, the Japanese began to teach TPM strategies internationally and were recognized for their results.

This process was an evolutionary one that took time, not because it was technically difficult to produce the results, but because of the efforts to change the organizational culture so that it valued the "Total" concept.

Today the international focus on TPM is intensifying. This interest is expressed to support a company's full utilization of its assets. For example, one of the prevalent strategies today is the concept of Lean Manufacturing. It is based on the Toyota production system and is designed to drive out waste from an organization. Lean Manufacturing strategies have yet to produce the true benefits possible because they assume full asset utilization. Furthermore, the full utilization of assets will never occur without an effective TPM strategy. Therefore, are Lean Manufacturing stra-tegies effective today? The answer is no. A quick review of the current state of maintenance in the United States indicates that changes are required if companies want to achieve the benefits of Lean Manufacturing.

Maintenance Costs

Various financial studies showed U.S. companies were spending over 600 billion dollars on maintenance and related expenditures in 1990. Of this huge amount, approximately one-third was unnecessary or wasted. This waste provides a cost advantage that companies can ill afford to give to their international competitors.

Where are the wastes? They are in the ineffective use and control of maintenance resources, labor, and materials. For example, what is the percentage of time that a maintenance technician is involved in actual hands-on activities? Is it two hours out of eight? Three hours? In companies where reactive or emergency types of maintenance make up 50% or more of the maintenance workload, technicians average only 2-3 hours of hands-on activities per day. During the rest of their time, they are engaged in non-productive activities such as looking for parts, drawings, instructions, or authorization.

What about inventory wastes? The cost of having too many spares is paid, not only in capital investments, but also in carrying costs, storage costs and labor costs. Still other costs include spoilage costs, pilferage costs, and the costs of damage caused by materials being stored and moved frequently.

A recent survey of maintenance and maintenance-related personnel showed organizational issues that were impacting maintenance efficiency and effectiveness. These areas include:

    Maintenance scheduling
    Hiring and training maintenance technicians
    Too much emergency or breakdown maintenance
    Lack of controls over maintenance spares
    Lack of upper management support and understanding

Each of these problems are difficult to solve, but when combined provide any manager with a formidable task. However, organizations that have these problems will have an almost impossible task trying to implement a TPM program. The right step is to solve some of these basic problems first, before tackling the task of implementing TPM. Later in the text, a methodology is presented showing how to solve these problems.

Maintenance Budgets

Maintenance budgeting is also another problem for organizations. Many methods are used to budget and monitor maintenance. While some work well, others are burdens to the maintenance departments. An extreme case occurs when maintenance is responsible for all maintenance-incurred expenses, whether these expenses are requested or approved by maintenance or operations. In such cases, maintenance may or may not be in control of the moneys be-

8 TOTAL PRODUCTIVE MAINTENANCE

ing spent. At the opposite end of the spectrum is the organization with a zero maintenance budget. All charges, including an overhead multiplier, are billed directly to the department requesting the work or which is owns the equipment. In this case, the operations or facility organization will want to keep the maintenance costs as low as possible, and will defer repairs, improvements, and refurbishes that should have been performed. In either case, the primary consideration, which is the condition of the company's capital assets (the equipment or facility), is pushed into the background. Total Productive Maintenance's focus on the equipment and / facility pays tremendous benefits to the company.

Figure 1-1 highlights the cost and payback for maintenance verses the company cost. Maintenance costs are between 15% and 40% of the total cost of production in typical manufacturing. The average is approximately 28%, an amount that is too high. When maintenance costs are reduced, even by as little as 10%, any cost avoidance is transformed directly into pre-tax profit. Some companies are able to save as much as 50% of their maintenance budget without sacrificing efficiency or quality of the maintenance work completed. Such savings, which increase pre-tax profit considerably, allow companies to be even more competitive in their respective markets. The comparative savings is highlighted in Figure 1-2.

These savings are enough of an enticement for some companies, but the true cost savings is yet to come. Consider which is

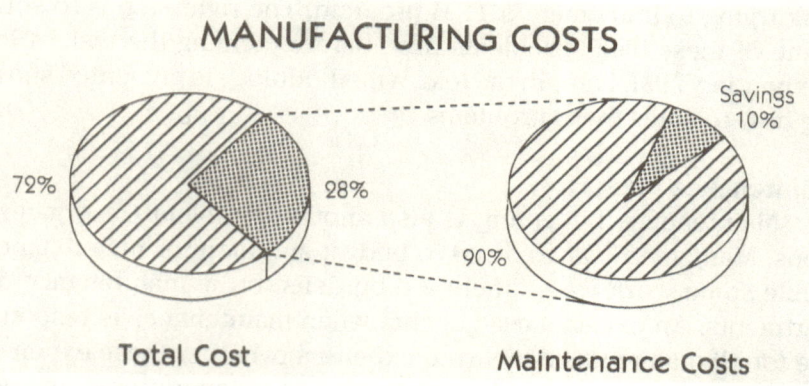

Figure 1-1 Reducing maintenance costs by 10% can produce an increase in pre- tax profit of almost 36%.

THE HISTORY AND IMPACT OF TPM

more: The maintenance cost of a repair or the cost of lost production? One survey showed that this costs ranges from 2:1 to as high as 15:1, as shown in Figure 1-3. Therefore, if the maintenance cost for a repair is $10,000, the true cost to the company of not having the maintenance work performed ranges from $30,000 to $160,000. It is critical that companies examine the true cost of maintenance versus non-maintenance if they are ever to be successful in improving maintenance and implementing TPM.

An additional problem is the control of the maintenance budgeting process. In over half of the sites, the plant manager or the plant engineer controls the maintenance budget, preventing the maintenance managers from controlling the departments they are responsible and held accountable for managing (see Figure 1-4). Unless managers are allowed to control their department budgets, they cannot be responsible for effectiveness. Total Productive Maintenance places responsibility and control for the job functions with the correct managers.

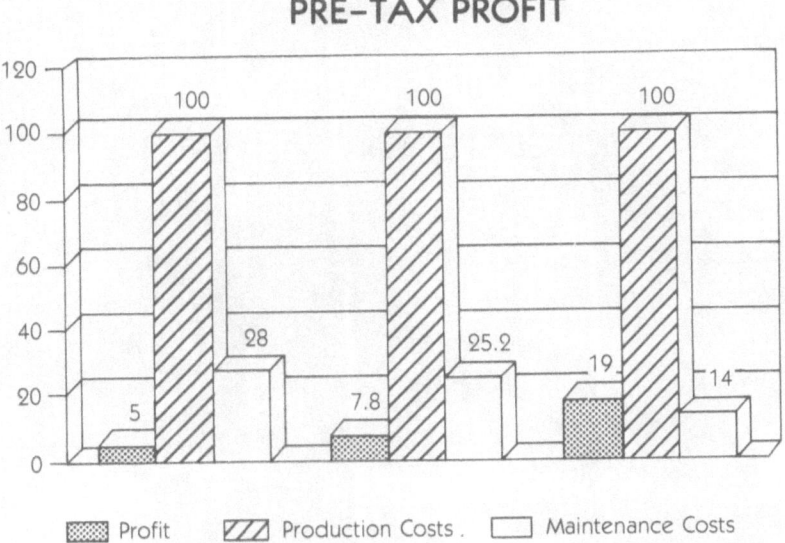

Figure 1-2 The relationship between reduced maintenance costs an increase in pre-tax profit.

10   TOTAL PRODUCTIVE MAINTENANCE

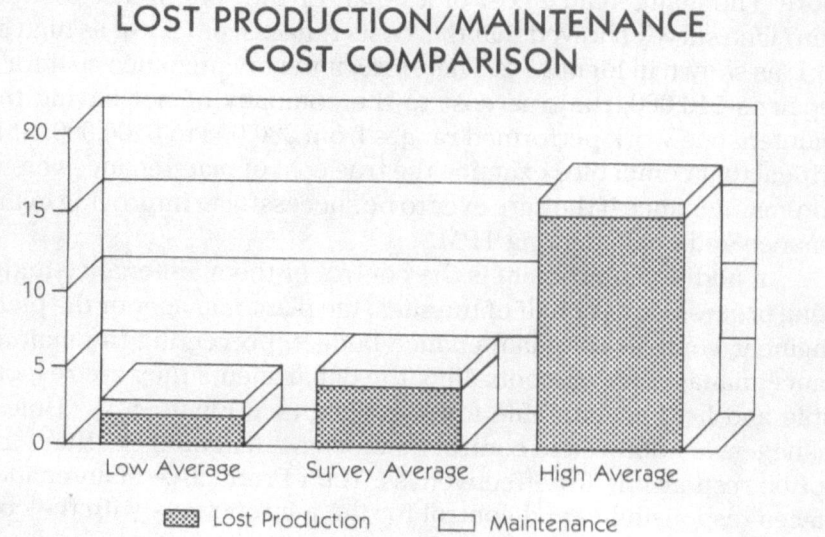

Figure 1-3 Lost production costs resulting from a maintenance repair can vary greatly in different companies.

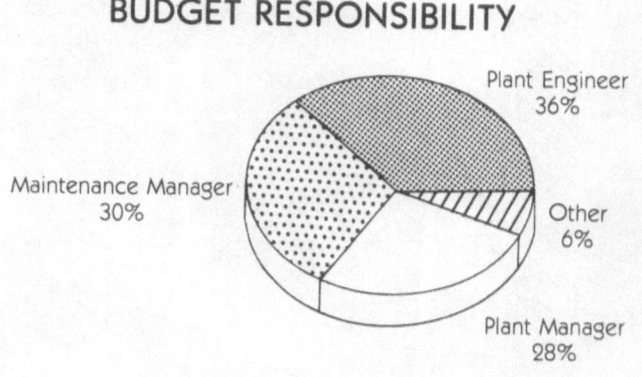

Figure 1-4 In only 30% if all sites does the maintenance manager control the maintenance budgeting process.

THE HISTORY AND IMPACT OF TPM

Maintenance Control Systems

This problem area contributes to all aspects of maintenance and is a key component of a successful TPM program. It is the system used by maintenance to gather information and provide an engineering database to make accurate and cost effective maintenance decisions. The more common name for this system is the work order. Most companies claim to have a work order system. However, only a minority of the companies are satisfied with their information, a point highlighted in Figure 1-5, which shows that the basic information-gathering function in most maintenance organizations is not functioning properly. If the information is not being gathered properly, one must question the accuracy of the decisions that are being made based on this information.

Companies that accurately gather information on work order systems still fail to use it correctly, showing a lack of performance monitoring and information analysis (see Figure 1-6). Thus, even when companies do gather the information, they fail to use it to find and implement cost effective asset management decisions.

Beyond the asset management issue, consider maintenance staff sizes. The size of a maintenance workforce is determined by the amount of work that it has to perform. This amount, commonly called the craft backlog, is the accumulated total of all estimated la-

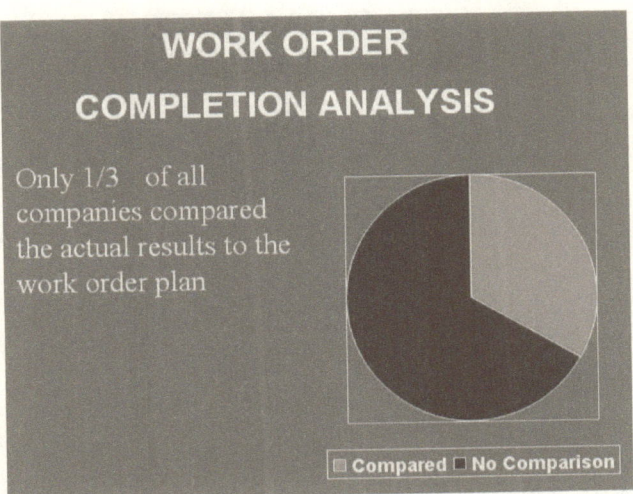

Figure 1-5 Work order system analysis.

12 TOTAL PRODUCTIVE MAINTENANCE

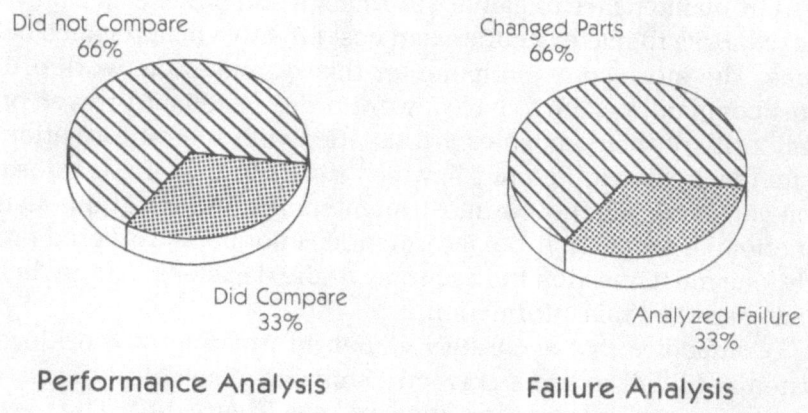

Figure 1-6 Even though almost 77% of all companies use work orders, many of them fail to analyze the information they provide.

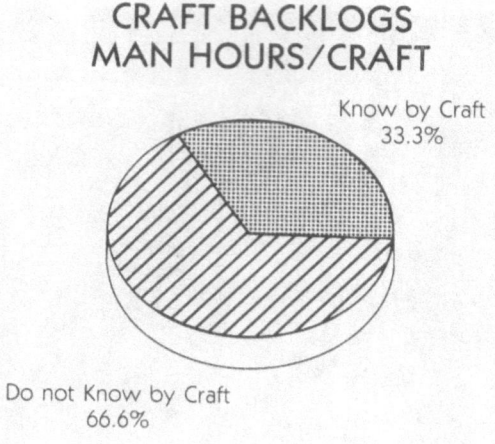

Figure 1-7 Many companies fail to plan and estimate existing work orders.

THE HISTORY AND IMPACT OF TPM   13

bor requirements on work orders waiting to be performed. Because many companies do not plan and estimate work orders, they do not know the size of their maintenance backlogs (see Figure 1-7). How do they then make justifiable decisions on maintenance staffing levels? It would be unimaginable for a production department to be managed in the same manner. You would either have operators standing around or equipment setting idle without an operator. This area must be controlled if optimum use of resources is to be realized.

Preventive Maintenance

Preventive maintenance is another major area that must be investigated during a TPM development plan. How do U.S. companies perform preventive maintenance? Almost 80% of the companies are not satisfied that their programs work or are cost effective (see Figure 1-8). The main reasons for these failures and the solutions will be discussed in Chapter 6. However, the largest reason for the failure of preventive and predictive maintenance programs is the lack of understanding and support for the program by upper management. Total Productive Maintenance programs ensure this support. Preventive maintenance under a TPM program will be successful if they are properly designed and implemented.

Maintenance Inventory and Purchasing

This area is a major contributor to the lack of maintenance productivity in the United States. The time wasted while trying to find parts for maintenance technicians makes up one of the largest por-

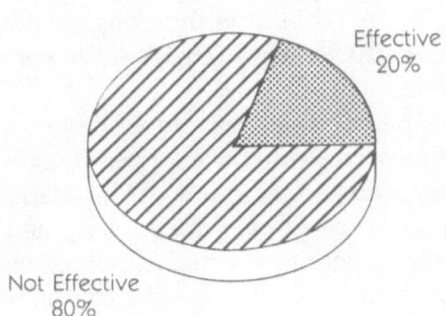

Figure 1-8 About 80% of all companies are not satisified with their existing preventive maintenance programs.

tions of the lost productive time already discussed. Estimates suggest that over 50% of all lost maintenance productivity is related to inventory and purchasing practices.

This problem is even further compounded by the fact that in almost half of all companies, maintenance has no control, or even input, over its inventory and purchasing policies. Therefore, order policies and storage policies are made by individuals who may not understand how maintenance inventory is different from operations or production inventory. This lack of understanding creates stockouts and overstocks, both of which are unnecessary expenses that weaken an organizations competitive position. While a stockout is considered a nuisance to maintenance, what is its true cost? What is the cost of downtime or lost production that is caused by a stockout? This cost can be considerable, yet it is rarely a factor considered in stocking decisions — a major flaw in many company's maintenance inventory and purchasing functions.

Is There Only One TPM Methodology?

Many companies are asking themselves if there is more than one way to approach TPM. Experts from around the world acknowledge problems implementing a cookbook-style TPM in any company due, in part, to factors such as:

    Skill of the workforce
    Age of the workforce
    Complexity of the equipment
    Age of the equipment
    Company culture
    Current status of the maintenance program

Remember that the primary focus of TPM is the constant improvement in the overall equipment effectiveness (OEE) of the company's equipment and capital assets.

The steps necessary to develop a TPM program must be determined for each company individually. These steps must be adjusted to fit individual requirements because the types of industry/service/facility, production methods, service activities, equipment conditions, special needs, problems, techniques, and levels of sophistication of maintenance vary dramatically from organization to organization.

THE HISTORY AND IMPACT OF TPM

Thus, there is no one right answer. The focus of any TPM program is the achievement of the five basic goals discussed earlier in this chapter as well as any others that may be identified by your particular organization. Although there is no one fixed methodology for implementing TPM, some guidelines can be examined from a technical perspective. These guidelines are presented in a flowchart format in the next section.

TPM Decision Tree

Good, sound maintenance practices are essential for effective TPM. But what exactly are good, sound maintenance practices?" The following discussion explains each block of the TPM decision tree (see Figure 1-9) designed to assist in the development of a TPM implementation methodology. These activities are not intended to be a cookbook, but rather to serve as a general model. If a clear understanding of the model is developed before attempting to implement TPM, then the implementation process will take less time and fewer resources.

1. PM Program Development.

Preventive maintenance is the core of any equipment maintenance process improvement strategy. All plant equipment, including special back up or redundant equipment, must be covered by a complete, cost-effective preventive maintenance program. The preventive maintenance program is designed to eliminate all un-planned equipment failures.

2. Evaluate the PM Program.

Evaluating the preventive maintenance program insures proper coverage of the critical equipment of the plant or facility. The program should include a good cross section of the following:
- Inspections
- Adjustments
- Lubrication
- Proactive replacements of worn components

The program should support the goal of no unplanned equipment downtime. This topic will be discussed further in Chapter 6.

3. Is it Effective? Less Than 20% Reactive Work Performed

The effectiveness of the preventive maintenance program is determined by the level of unplanned equipment maintenance that is performed. Unplanned equipment maintenance is defined as any maintenance activity that is performed with less than one week of advanced planning. Unplanned equipment maintenance is commonly referred to as reactive maintenance.

An effective preventive maintenance program will reduce the amount of unplanned work to less than 80% of the total labor expended for all equipment maintenance activities. If more time is being spent on unplanned activities, then a reevaluation of the preventive maintenance program is required. It will be difficult to make progress in any of the following areas unless the preventive maintenance program is effective enough for the equipment maintenance to meet the 80% / 20% rule.

4. Review the Maintenance Stores.

Once the preventive maintenance program is effective, the equipment spares, inventory, and purchasing systems must be analyzed. The equipment spares and inventory should be organized, with all of the spares identified and tagged, then stored in an identified location, with accurate on-hand and usage data. The purchasing system must allow for procurement of all necessary spare parts to meet the maintenance schedules. All data necessary to track the cost and usage of all spare parts must be complete and accurate. This topic will be discussed further in Chapter 7.

5. Are the Stores Effective?
(Greater Than 95% Service Level)

Simply defined, the service level measures the percent of time that a part or in stock when it is requested. The spare parts must be on hand at least 95% of the time for the stores and purchasing systems to support equipment maintenance activities. Unless maintenance activities are proactive (less than 20% unplanned weekly), the stores and purchasing groups cannot be cost effective in meeting equipment maintenance spare parts demands. However, the inventory and purchasing functions must achieve at least a 95% service level before effective work order utilization can occur.

6. Review the Work Order System.

The work order system is designed to track all equipment maintenance activities. The activities can be anything from inspections and adjustments to major overhauls. Any maintenance that is performed without being recorded in the work order system will be lost. In turn, lost or unrecorded data makes it impossible to perform any analysis of equipment problems. All activities performed on equipment must be recorded to a work order by the responsible individual. This step highlights the point that maintenance, operations, and engineering will be extremely involved in utilizing work orders.

7. Are Work Orders Fully Utilized? (100% Coverage)

This question should be answered by performing an evaluation of the equipment maintenance data. The evaluation can be as simple as answering the following questions:

How complete is the data?
How accurate is the data?
How timely is the data?
How usable is the data?

If the data is not complete, it will be impossible to perform any meaningful analysis of the equipment's historical and current condition. If the data is not accurate, it will be impossible to correctly identify the root cause of any equipment problems. If the data is not timely, then it is impossible to correct equipment problems before they cause equipment failures. If the data is not usable, it will be impossible to format it in a manner that allows for any meaningful analysis. Unless the work order system provides data that passes this evaluation, further progress will be impossible.

For example, consider team problem-solving activities that are focused on improving overall equipment effectiveness. The teams always look for the root causes of problems that impact the OEE. Without accurate data from the work order, how can they perform a root cause analysis? How can they identify the top ten problems that cause downtime on the equipment? How would they know which modifications have been done to the equipment in the past that could have caused the current problem? Without data, all decisions about the equipment become subjective.

8. Review Planning and Scheduling.

This review examines the planning and scheduling policies and practices for equipment maintenance. The goal of planning and scheduling is to optimize any resources expended on equipment maintenance activities, while minimizing the interruption the activities have on the production schedule. The goal of planning and scheduling is to insure that all equipment maintenance activities occur like a pit stop in a NASCAR race. This insures optimum equipment uptime, with quality equipment maintenance activities being performed. Planning and scheduling pulls together all of the activities (maintenance, operations, and engineering), and focuses them on obtaining maximum quality results in a minimum amount of time.

9. Are Planning and Scheduling Effective? (Greater Than 80% Weekly)

While this step is similar to Step 3, its focus is on the effectiveness of the activities performed in the 80% planned mode. An effective planning and scheduling program will insure maximum productivity from those employees performing any equipment maintenance activities. Delays, such as looking for parts, rental equipment, drawings, or tools, or waiting while equipment is shut down, will all be eliminated. If these delays are not eliminated through planning and scheduling, then optimizing equipment utilization will be impossible. It will be equivalent to a NASCAR pit crew taking too long for a pit stop; the race is lost by not keeping the car on the track. Equipment utilization is lost by not properly keeping the equipment in service. Maintenance planning and scheduling is discussed further in Chapter 8.

10. Investigate the Computerization of the Work Order System.

A considerable volume of data is generated and tracked to properly utilize the work order system and to plan and schedule effectively. If the data becomes difficult to manage using manual methods, it may be necessary to computerize the work order system. If the workforce is burdened with excessive paper work or is accumulating file cabinets of equipment data that no one has time to look at, then it is best to computerize the work order system. However, if the number of pieces of equipment is relatively small and data tracking and analysis are not a burden, then it may be best to maintain the manual work order system.

THE HISTORY AND IMPACT OF TPM   19

10a. Establish a Manual Equipment Maintenance System.

A manual system can be as simple as a cardex file with cards for each equipment item, and with notations of all repairs and services on the cards. Other methods include a visual white board with markers and spaces for notations or a magnetic board with tags that can be moved as each service is complete. Still another method is a log book, which may simply be a three-ring binder, with pages for notations of each service or repair that is performed on the equipment. It does not matter which method is used, but rather that the equipment data is complete and in a format that can be analyzed.

10b. Is the Manual System Effective?

The manual system should meet the equipment management information requirements of the organization. Some of these re-quirements include:
  Complete tracking of all repairs and service
  The ability to develop reports, for example:
    Top ten equipment problems
    Most costly equipment to maintain
    Percent reactive vs. proactive maintenance
    Cost tracking of all parts and costs

If the manual system does not produce this level of data, then it needs to be re-evaluated. (If it is effective, then go to Step 13.)

10c. Evaluate the Manual Work Order Process.

The goal of reevaluating the manual work order system is to determine where the weaknesses are in the system so that they can be corrected and good equipment data can be collected. Several questions for consideration include:
  Is the data we are collecting complete and accurate?
  Is the data collection effort burdening the work force?
  Do we need to change the methods we use to manage the data?
  Do we need to re-evaluate the computerization decision?

Once problems are corrected and the equipment management information system is working, then constant monitoring for problems and solutions must be put into effect. (Go to Step 13.)

11. Purchase and Implement a CMMS (EMIS).

The computerized maintenance management system (CMMS), also known at times as an Equipment Management Information System (EMIS), is a computerized version of a manual system. There are currently over 200 commercially produced CMMSs in the North American market. Finding the correct one may take some time, but through the use of lists, surveys, and word of mouth, and by evaluating the vendor's financial status, it should take no more than three to six months for any organization to select a CMMS. Once the right CMMS is selected, it must then be implemented. CMMS implementation may take from three months for smaller organizations to as long as 18 months for larger organizations to implement.

Companies can spend much time and energy addressing CMMS selection and implementation. Keep in mind that CMMS is only a tool to be used in the improvement process; it is not the goal of the process. Losing sight of this fact can curtail the effectiveness of any organization's path to continuous improvement.

12. Is the CMMS Usage Effective?

If the correct CMMS is selected, then it makes the equipment data collection faster and easier. It should also make the analysis of the data faster and easier. The CMMS should assist in enforcing World Class maintenance disciplines, such as planning and scheduling and effective stores controls. The CMMS should provide employees with usable data with which to make equipment management decisions. If the CMMS is not improving these efforts, then the effective usage of the CMMS needs to be evaluated. Some of the problems encountered with CMMS include:

Failure to fully implement the CMMS
Incomplete utilization of the CMMS
Inaccurate data input into the CMMS
Failure to use the data once it is in the CMMS

CMMS will receive further consideration in Chapter 9.

13. Investigate Operator Involvement.

As the equipment management system (CMMS, EMIS) becomes effective, it is time to investigate whether operator involve-

THE HISTORY AND IMPACT OF TPM

ment is possible in some of the equipment management activities. There are many issues that need to be explored, including the types of equipment being operated, the operators to equipment ratios, the skill levels of the operators, and contractual issues with the employees' union. In most cases, there is some level of activity at which the operators can be involved within their areas. If there are no obvious activities for operator involvement, then a reevaluation of the activities will be necessary.

14. Identify the Activities.

The activities in which the operators may be involved can be either basic or complex. The complexity is determined partially by their current operational job requirements. Some of the more common tasks for operators include, but are not necessarily limited to:

A. Equipment Cleaning

This activity may be as simple as wiping off the equipment when starting it up or shutting it down.

B. Equipment Inspecting

This activity may range from a visual inspection while wiping down equipment to a maintenance inspections checklist utilized while making operational checks.

C. Initiating Work Requests

Operators may prepare work requests for any problems (either current or developing) on their equipment. They pass these requests on to maintenance for entry into the work order system. Some operators will directly input work requests into a CMMS.

D. Visual Systems

Operators may use visual control techniques to inspect their equipment and to make it easier determine its condition.

Whatever the level, operator involvement should contribute to the improvement of the equipment effectiveness.

15. Are the Operators Certified to Perform the Activities?

Once the activities in which the operations personnel are to be involved have been determined, the operators' skills to perform these activities need to be evaluated. The operators should be properly trained to perform any assigned tasks. The training should be developed in both written and visual formats. Once the operators are trained, copies of the materials should be given to the operators for their future reference. These materials will contribute to the commonality required in order for operators to be effective while performing these tasks. In addition, certain regulatory organizations require documented and certified training for all employees (e.g., Lock Out Tag Out). Training for personnel involved in TPM will be discussed further in chapter 10.

16. Begin Operator Involvement.

Once the operators are trained and certified, they can begin performing their newly-assigned tasks. The operators must be coached for a short time to insure they have the full understanding of all aspects of the new tasks. Some companies have made this coaching more effective by having the maintenance personnel assist with it. Background knowledge can then be transferred to the operators—information that they may not have received otherwise during the more formal training.

17. Is Predictive Maintenance Being Performed?

Once the operators have begun performing some of their new tasks, maintenance resources may become available for other activities. One area that should be explored is predictive maintenance. Fundamental predictive maintenance techniques include:

Vibration Analysis
Oil Analysis
Thermography
Sonics

Plant equipment should be examined to see if any of these techniques will help reduce downtime and improve service. Pre-dictive technologies should not be utilized because they are technically advanced, but only when they contribute to improving the equipment effectiveness. The correct technology should be used to trend or solve the equipment problems encountered.

THE HISTORY AND IMPACT OF TPM   23

18. Investigate Reliability Engineering.

Reliability Engineering is a broad term that includes many engineering tools and techniques. Some common tools are:

Life Cycle Costing. This technique allows companies to know the cost of their equipment from when it was designed to the time of disposal.

RCM. Reliability Centered Maintenance is used to track the types of maintenance activities performed on equipment to insure that they are the correct activities to be performed.

FEMA. Failure and Effects Mode Analysis examines the way the equipment is operated as well as any failures incurred during the operation in order to find methods of eliminating or reducing the numbers of failures in the future.

Early Equipment Management and Design. This technique takes information on equipment and feeds it back into the design process to insure that any new equipment is designed for maintainability and operability.

Using these and other reliability engineering techniques improves equipment performance and reliability and, in turn, helps to insure competitiveness.

19. Investigate Financial Optimization.

Once the equipment is correctly engineered, the next step is to understand how the equipment or process impacts the financial aspects of the company's business. Financial optimization considers all costs impacted when equipment decisions are made. For example, when calculating the timing to perform a preventive maintenance task, are the cost of lost production or downtime considered? Are wasted energy costs considered when cleaning heat ex-changers or coolers? In this step, the equipment data collected by the company are examined in the context of the financial impact they have on the company's profitability.

20. Are the Tools and the Data Available for Financial Optimization?

While financial optimization is not a new technique, most companies do not properly utilize it because they do not have the data necessary to make the technique effective. Some of the data required includes:

MTBF (Mean Time Between Failure) for the equipment
MTTR (Mean Time To Repair) for the equipment
Downtime or lost production costs per hour
A Pareto of the failure causes for the equipment
Initial cost of the equipment
Replacement costs for the equipment
Complete and accurate work order history for the equipment

Without this data, financial optimization cannot be properly conducted on equipment. Without the information systems in place to collect this data, a company will never have the accurate data necessary to perform financial optimization. We will look more closely at financial optimization in Chapter 11.

21. Use Financial Optimization.

If the data exists and the information systems are in place to continue to collect the data, then financial optimization should be utilized. With this tool, equipment teams will be able to financially manage their equipment and processes.

22. Evaluate the Success of the TPM Program.

Are the results achieved by maintenance reaching the goals that were set for the improvement program when it was started? If not, then the maintenance improvement program needs to be examined for gaps in performance or deficiencies in existing parts of the process. Once weaknesses are found, then steps should be taken to correct or improve these areas.

23. Strive for Continuous Improvement.

Continuous improvement means never getting complacent. It calls for constant self-examination with the focus on how to become the best in the world at the company's business.

This implementation flow focuses on the "technical" flow to TPM. Meanwhile, there is a complementary flow chart that examines the "people" side of TPM. This flow chart, developed by Robert Williamson of Strategic Work Systems, is available by contacting him at www.swspitcrew.com

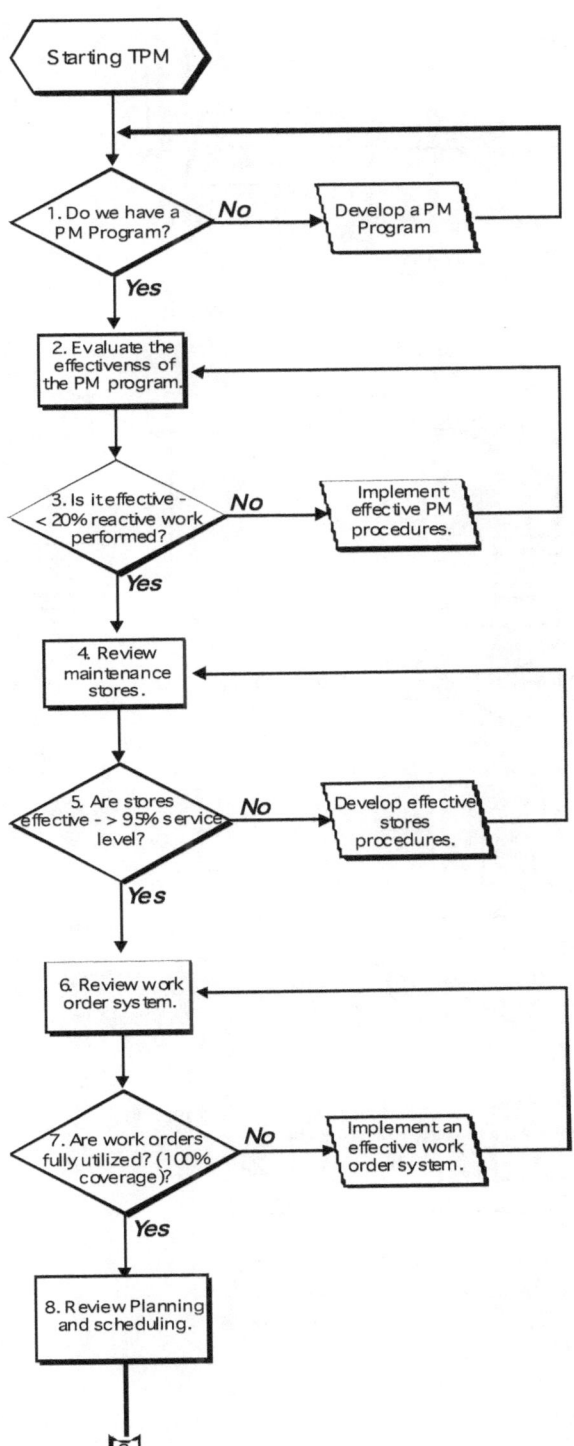

Figure 1-9 a

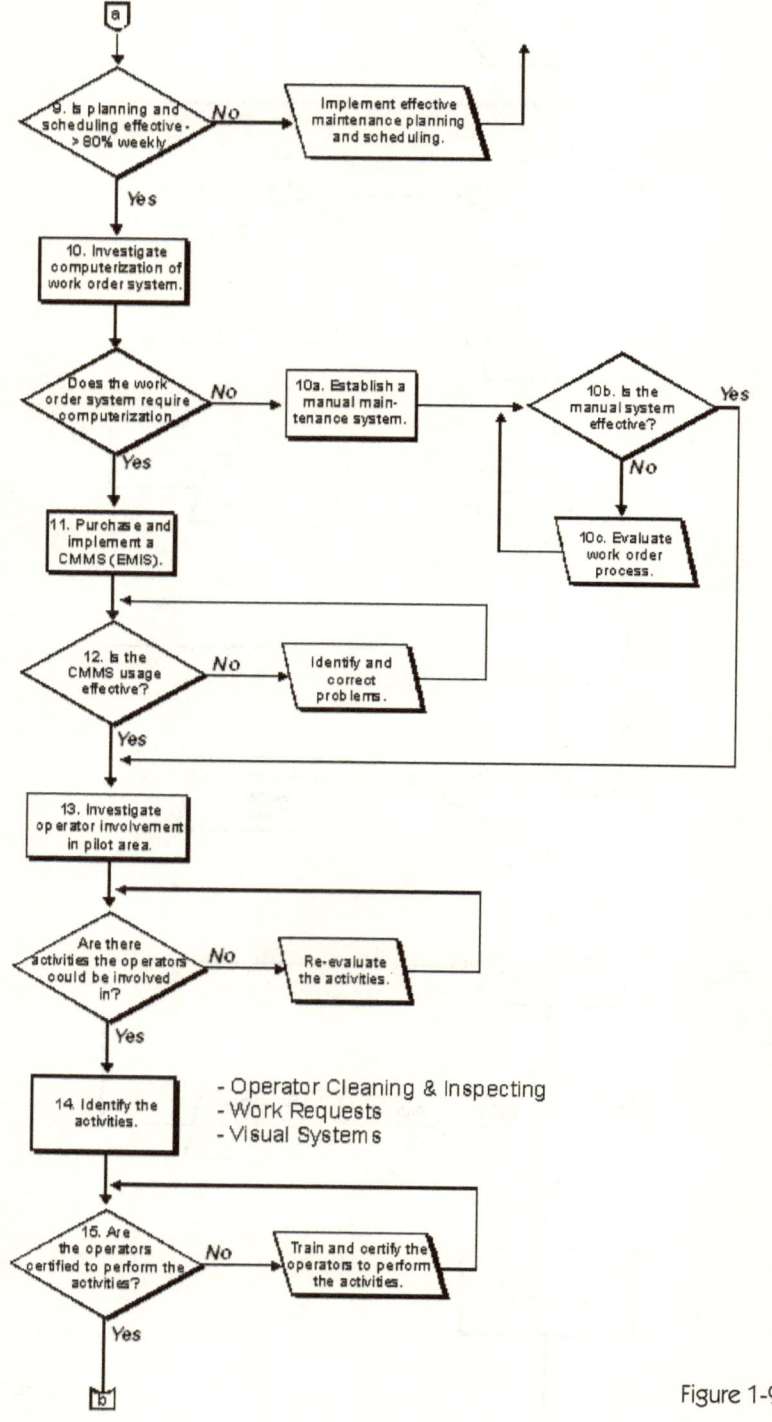

Figure 1-9 b

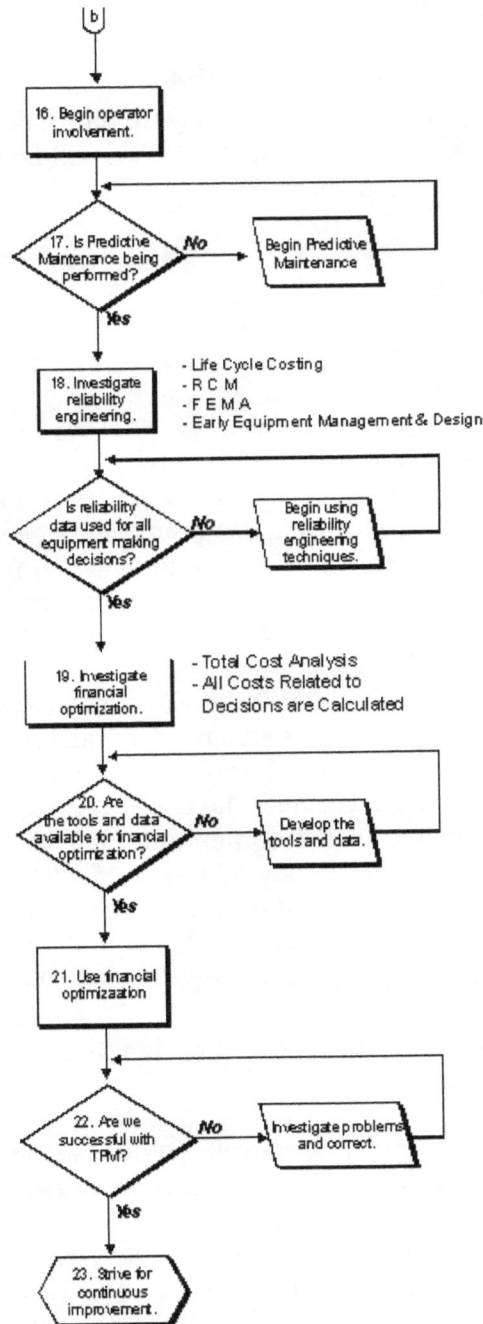

Figure 1-9 c

Pitfalls for TPM

While the TPM flow appears to be easy to follow, there are hidden pitfalls when trying to implement TPM. Two will be considered now, with these and others receiving additional consideration in Chapter 15.

Organizational Downsizing

In the business environment of the early 2000s, much of management focus is spent on headcount reductions or downsizing of the workforce. This practice is detrimental to the employee involvement required by TPM. In some companies, TPM starts as a middle management activity. The line employees buy in to the common sense approach to TPM and begin to contribute ideas that increase productivity. Because senior management has never been properly educated about the process, they use the increase in productivity (output) to focus on reducing expenses to further increase profits. When this occurs, the employee involvement required by TPM diminishes and the TPM strategy fails.

In the January 14, 1995 issue of the Economist, it was stated that, "Even the 1980s' most ardent adherents of quality are finding that TQM does not readily blend with wave after wave of restructuring, downsizing, and re-engineering". The article stated further "the snag is downsizing undermines a cornerstone of TQM: Employee Motivation." If you change the acronym from TQM to TPM, you have an equally true statement. Unless the entire organization from senior management to the line employees understand the true focus of TPM—which is improving equipment effectiveness—the TPM effort will be destined to failure.

Senior executives today need to realize that there are two ways to decrease cost per unit.

1. Make the same number of units and reduce expenses.
2. Hold the line on expenses and make more units.

TPM, while doing both, focuses more on increasing equipment utilization and eliminating waste (reducing expenses) rather than focusing on headcount reductions. Unless this is clearly understood throughout the organization, companies will continue to struggle to implement TPM strategies.

Lack of Focus on Results

Many consulting organizations that teach TPM focus on its esoteric aspects. They have their clients start by cleaning their equipment, forming teams to discuss theoretical improvements, and creating visual systems to make the plant look better. Although these activities are a part of the overall TPM strategy, they are implemented without any tangible results. Therefore, the companies spend their physical and financial resources with little, if any, financial return on investment.

Unless all of the initiatives in TPM are tied to financial benefits or improvements, senior management support wanes over time. When the company has a particularly tight quarter financially, senior management eliminates support for strategies that are not providing an immediate return on investment. When this occurs, the plug is too easily pulled on the TPM strategies. However, in companies where all TPM initiatives are tracked and financially justified, the overall TPM strategy is supported by senior management even in the most difficult of times.

TPM is not a difficult strategy to execute. As long as it has a focus and financial justification, it will be successful. This theme will continue to be developed throughout the remainder of this text.

CHAPTER 2

DEFINING EQUIPMENT LOSSES

The true goal of TPM focuses on the utilization of corporate assets. This includes the condition of equipment, the buildings and grounds, and the entire facility. Of TPM's five basic goals, repeated below, improving equipment effectiveness is the most important.
1. Improving equipment effectiveness
2. Improving maintenance efficiency and effectiveness
3. Early equipment management and maintenance prevention
4. Training to improve the skills of all people involved
5. Involve operators (occupants) in routine maintenance

Improving equipment effectiveness insures that equipment operates at its design specifications for the entire life of the equipment. The goal for the equipment is to increase its effectiveness or capabilities by minimizing input and maximizing output. This goal applies to facilities as well as to production plants.

In facilities, the input includes the labor resources, energy resources, supplies, and other miscellaneous resources that are utilized at the site. For example, unless the physical plant and building heating, lighting, etc. are properly maintained, there is waste. Energy usage is higher than required, wasting a corporate asset: money. The building should be maintained at a level where the maximum benefit is obtained from each resource used by the facility.

In production environments, the input includes the resources required to produce the product. The output is:
Increased capacity and productivity
Highest quality
Lower production costs
On-time delivery
Improved safety and hygiene
Higher employee morale
A more favorable workplace environment

Obstacles To Success: Attitude

These benefits are desirable to any company, but there are obstacles in the path to achieving these goals. The first major obstacle is the attitude of all personnel, from top management to the line employees, one that is willing to accept anything less than the design performance specification for the equipment. Too many employees at all levels accept the Murphy's Law suggesting that "All equipment is just waiting to break down". In quality assurance programs, employees are taught to accept zero defects. In equipment and facility maintenance, they should be taught to expect zero breakdowns and zero failures. Although breakdown losses are readily recognized, there are several significant causes of poor equipment utilization. They are:
1. Breakdown losses
2. Setup and adjustment losses
3. Idling and minor stoppages
4. Reduced capacity losses
5. Quality problems
6. Startup/Restart losses
7. Other related costs

1.0 Breakdown Losses

Two major types of breakdown losses must be recognized: capacity loss and capacity reduction breakdowns. Capacity loss is the easiest to recognize because the equipment ceases to operate. This loss, typically called a breakdown, is the production stoppage or the stoppage of a service to a facility. The maintenance department responds in an emergency mode and works to quickly restore the operation or service.

Capacity reduction breakdown is more subtle in nature. As the equipment ages, it experiences wear. As the wear continues, the capacity of the equipment begins to decline. Unless careful monitoring occurs, the reduced capacity goes unnoticed or is accepted as normal. In production, this translates into slower operation, lower capacities, and increased labor costs. In facilities, it translates into higher energy and operational costs. For example, failure to maintain adjustments and calibrations on HVAC systems may result in a 25% increase in energy costs.

While capacity loss breakdown is the easiest to find and repair, the capacity reduction breakdown represents the largest cost to most corporations. In the majority of cases, capacity loss breakdown is a technical problem whereas capacity reduction breakdown is an organizational problem.

Capacity loss breakdown is caused by the failure of some component on the equipment. Because most preventive and predictive maintenance programs are designed to detect and trend normal wear, other types of wear will cause the majority of these breakdowns. These other types include infant mortality failures, random failures, and failures related to poor operating and maintenance activities. If the preventive or predictive maintenance program is effective, these breakdowns will be minimal in nature.

Capacity reduction breakdown is generally caused by neglect of a chronic equipment problem. This problem occurs over a long period of time and becomes accepted as a normal fact of operation. A typical example is equipment that produces a defective product when operating at any rate over 80% of design speed. Instead of taking the time and effort necessary to correct the problem, the organization issues a memo stating not to run the equipment over 80% of design speed. This approach results in a 20% reduction in equipment capacity. If this process is repeated over several years, the plant will soon need to invest in new equipment just to meet the necessary production rate. The problem becomes severe because management is focusing on short-term goals rather than long-range goals.

When capacity reduction problems develop, solving the problem is more economical than reducing the operating standards. But management must examine short-term profits in the light of long-term profitability.

Eliminating both capacity reduction and loss breakdowns is included in the zero breakdown concept. If this goal is to be realistic, the company must be committed to a program designed to prevent breakdowns. The program must address the different problems, both organizational and technical that contributes to the breakdowns. Such a program will consist of five basic parts. They are:
1. Maintaining basic equipment condition
2. Maintaining designed operational standards
3. Restoring normal equipment wear
4. Reliability engineering
5. Eliminating human error

1.1 Maintaining Basic Equipment Conditions

Maintaining basic equipment conditions is a maintenance practice that is ignored in most companies today. The maintenance group is occupied on capacity loss breakdowns, trying to keep the equipment operating properly. Meanwhile, basic tasks like inspections, cleaning, lubricating, adjusting, and tightening are neglected. The resulting deterioration and wear caused by lack of service contribute to additional capacity loss breakdowns, resulting in less service. This snowball effect spells disaster for an organization.

In a true TPM organization, the operators take responsibility for some of these tasks, allowing maintenance personnel to focus on other tasks. Cleaning the equipment is critical because any contamination accelerates wear and increases deterioration. For example, contamination on a gearcase shell, usually caused by lubricant covered with dust, develops a barrier between the gearcase and the atmosphere. Because a gearcase shell acts to help dissipate heat developed inside the case, the effectiveness of this design function is reduced. This loss leads to increased temperature in the gearcase, reducing the viscosity of the lubricant. In turn, the higher temperatures and reduced viscosity from the lubricant result in accelerated wear and deterioration, increasing the probability of a capacity loss breakdown.

The previous example is just the tip of the iceberg when it comes to the problems that cleaning equipment can prevent. Similarly, basic tightening can eliminate problems related to looseness and vibration, either of which accelerates mechanical wear, contributing to deterioration of the equipment. A problem related to maintaining proper tightness of mechanical equipment is proper torque adjustments. Just as looseness is a problem, excessive tension applied to fasteners is also an invitation to problems. Excessive tension exceeds the designed elasticity of the material. Once these limits are exceeded, as in a bolt, it becomes impossible to maintain tightness. The bolt will continue to stretch, allowing looseness and resulting in mechanical wear. The reverse is also true. When insufficient torque is applied to a fastener, it quickly will loosen. The resulting vibration accelerates wear on the fastener, resulting in premature failure. The answer to these types of problem is proper training and standards for tightening and adjustment. This answer seems basic, yet how many maintenance technicians actually use torque wrenches to install fasteners? Very few, which is why we have this problem.

Lubrication is also a key ingredient to maintaining basic equipment conditions. Effective lubrication programs insure that the right lubricant is applied in the proper quantity to the correct application points. This program also requires training the operations and maintenance personnel involved. Lack of lubrication accelerates equipment wear. However, excessive lubrication can also damage equipment by increasing temperature levels as the lubricant churns, damaging the equipment and the lubricant. If seals are damaged and leaking, contamination can enter through the defective seal, contributing to increased wear of the equipment. If lubricants from different vendors are mixed without comparing interchange charts, chemical additives will react and raise or lower the lubricant viscosity, the lubricant may thin or coagulate, or the lubricant can become acidic or alkaloid, causing adverse reactions with the very equipment it was supposed to protect. An effective lubrication program will insure that operations and maintenance personnel are properly trained in lubrication basics. Regular lubrication points may be color coded indicating what type of lubricate is to be used. Charts should be provided to insure the proper amount of lubricant is applied.

The best method to insure that the proper standards are adhered to in the above areas is to document the maintenance program. Many companies already have this step completed with their preventive maintenance program. However, in order to provide sufficient coverage and detail, the service and inspection instructions should be comprehensive. The instructions should be detailed; the instruction "Add lubricant if necessary" is not sufficient. The amount and type should also be included, so that mistakes can not be made. These standards will insure that the basic equipment conditions are maintained.

Further highlighting the importance of these basics was a study performed by Engineer's Digest magazine, showing that on average 50% of all equipment breakdowns have a root cause related to the neglect of basics. The question remains: How does your organization perform the basics of maintenance?

1.2 Maintaining Designed Operational Standards

This component of the breakdown elimination program is developed from the original design specifications for the equipment. Its focus is that the equipment should operate at its rated speed and capacity throughout its entire life cycle. The two alternatives are to operate it below capacity or to operate it above its rated capacity. Operating the equipment below its rated capacity is a waste of capital assets. It may cause the company to invest in unnecessary equipment and floor space to produce its product. In a facility it may mean buying extra environmental control equipment (such as HVAC systems) to maintain the facility. This waste is expensive and will certainly impact a company's ability to be cost competitive. Operating the equipment above its rated capacity will overload it, resulting in excessive wear on the equipment and also its subsequent premature depreciation.

When a company determines the design specifications for each piece of equipment or system and operates at these specifications, it insures it is obtaining the maximum value from each asset.

Once the standards are established, they should be documented and distributed to operational and maintenance personnel. These standards will help them insure that the equipment is properly maintained and operated. Environmental conditions should be included in these standards, including temperature, humidity, and contamination. Insuring that the equipment is operated in the specified environment again helps provide the maximum return on investment for each asset. Failure to do so will, as previously highlighted, contribute to excessive wear and premature failure.

Another area to notice in the standards for companies that

move equipment regularly covers construction and installation. These standards help insure that the equipment is properly installed or stored each time it is moved. The standards here could include electrical specifications, proper piping specifications, and mechanical installation specifications. For example, improper physical installation of equipment could allow for excessive vibration which, in addition to causing wear, could be a contributing factor to quality problems. The proper attention to maintaining equipment operating standards is a major part of insuring equipment capacity.

1.3. Restoring Normal Equipment Wear

Restoring normal wear is a process that continues for the entire life of the equipment. One of the first steps in the process is to set standards for the equipment condition. These standards should be set to a level that will assure equipment capacity. Therefore, a good system of inspection and repair must be set in place.

Engineering calculations such as mean time between failures (MTBF) and mean time to repair (MTTR) should be used to help set repair frequencies. These calculations will provide information necessary to set maintenance replacement and rebuild intervals. These indicators coupled with good use of predictive maintenance technologies should prevent any capacity reduction conditions from developing to a level that affects operations.

As the maintenance becomes more controlled, standard repair policies and practices should be developed. These policies and practices are usually in the form of standing or repetitive work orders. The standardized repair policies should include detailed instructions: lists of parts required and estimated labor resources required. These instructions eliminate the uncertainty and lack of coordination found during many equipment rebuilds. They also provide the ability to schedule the work properly, with minimum disruptions to the production or operations groups.

A subset of standard repair procedures is the standardization of inventory items and their storage. Such standardization means that the repair parts will be consistent and they will be able to be located. Presently in many companies, the purchasing and inventory groups change suppliers or manufacturers so often that they continually need updated usage of interchange charts. The situation often arises in which parts that were supposed to be interchangeable do not match. This problem is usually not discovered until the repair is in progress. It contributes to delays in making the repair and lost production or use of the facility. Standardization of spare parts and spares procedures is one of the most important support areas to insuring restoration and preservation of equipment condition.

1.4 Reliability Engineering

Reliability engineering is used to correct equipment-related problems that cannot be addressed through normal maintenance or operator-based maintenance procedures. Reliability engineering is generally applied to design weaknesses. It can be used to strengthen parts, extending the life of the equipment. There are usually chronic problems that directly impact equipment capacity, including wear resistance and corrosion resistance. Eliminating these two problems will contribute major improvements to equipment life cycles.

Another main part of reliability engineering focuses on design or material changes to reduce stress and fatigue of materials. These changes can be achieved by more accurate machining of parts, changing materials the parts are made of, or changing the shape of parts or components. In addition, the accuracy with which some components are manufactured can be increased, contributing to less wear on the component and related systems. If used in conjunction with the other capacity assurance techniques, reliability engineering will assure optimum availability.

A pitfall that commonly prevents companies from being effective in this are is the lack of understanding of the true design life of equipment components. For example, the average life expectancy of a v-belt is 24,000 hours of operation. This is approximately three years of life in a 24-hour-a-day – 7-days-a-week operation. However, many companies are pleased with themselves if they get six months to a year out of a v-belt. When a v-belt is installed, how many technicians really follow all of the steps of proper installation? These steps include:

1. Reduce the tension from the old belts.
2. Remove the old belts, inspecting for wear patterns.
3. Clean the sheaves with a brush to remove contamination.
4. Using a sheave gauge, check the sheave for excessive wear.
5. Using a belt gauge, check the new belts for proper size.

6. Lay the new belts in the sheaves.
7. Tighten the new belts.
8. Check the tension on the new belts with a tension gauge.
9. Check the alignment of the sheaves using a straightedge and string.
10. Restart the drive.
11. Recheck the tension after the belt has been in service a few hours with the tensioning tool.

In many cases, the old belts are cut off, the new belts run on the sheaves, the alignment and tension are not properly checked, and the belt attains only a fraction of its design life. These problems result in unnecessary maintenance costs, both from a labor and material perspective. If companies spent time training their technicians on proper care of basic equipment components, they would have the opportunity to make major reductions in their equipment maintenance budgets.

1.5 Eliminating Human Error

This equipment capacity problem has two basic divisions: operations and maintenance. The human error for operations is misuse or abuse of the equipment or facility. The human error for maintenance is lack of maintenance or repair errors.

Operational misuse or abuse of equipment is controlled by training. To solve the problem, the cause of the misuse or abuse must first be studied. If it is lack of knowledge on the part of the operators, they should be properly trained. If it is due to poor operating procedures, these must be updated and distributed. If it is a human-machine interface problem, then the design of the equipment interface must be improved. In some cases, the Japanese use the poka-yoka or fool-proofing technique to prevent misoperation of the equipment. This technique requires some simple device or alteration to the equipment that prevents any mistakes by the operator.

Maintenance errors will either be a failure to perform some maintenance act or an actual repair error. In either case, the solution is based on examining the causes of the problems and correcting the problem. For example, if the problem is a missed maintenance act, such as a preventive maintenance service, the cause for missing the PM should be investigated. If the cause was a conflict between the

DEFINING EQUIPMENT LOSSES

maintenance and operations schedule, then both groups must work toward a resolution; otherwise, the corporate assets are affected.

Additional maintenance errors may be corrected by improving the working conditions, the tools, and the equipment that maintenance utilizes in the repairs. Long-standing chronic problems may be solved by simplifying maintenance repair procedures and troubleshooting processes.

Eliminating the human factor in equipment capacity assurance programs will remove one of the most difficult to detect chronic problems.

2.0 Setup and Adjustment Losses

Setup and adjustment losses are incurred when equipment is used to manufacture different products. The definition of setup and adjustment time is the period of time lost when the equipment is finished producing one product and starts producing the next product at the specified quality. In some operations, this time period can be considerable. The losses incurred during this time can make a difference between a profit and loss for a company. For example, if the setup and adjustment time is reduced, the lot size that is economical to run can be reduced. This reduction increases the flexibility of the production scheduling and sales group, allowing them to be more responsive to the customer's requests. Responsiveness to the customer is a measure of the company's competitiveness and may make the difference between winning and losing the business.

Improvements in equipment setups and adjustments are typically divided into three areas:
1. External Setups. These activities can take place while the equipment is in operation.
2. Internal Setups. These activities can take place only while the equipment is shut down.
3. Adjustments. These activities take place while trying to get the equipment produce a quality product.

Each area can make a major contribution to the overall improvement.

2.1 External Setups

In order to be most effective with external setups, all assembly and adjustments must be made before the unit is taken out to the equipment to be installed. This step may involve preparing all tools and equipment that will be required to make the installation. The external part will also insure that all related tools are ready, such as precision measuring instruments. All measurements that can be set independent of the unit should be made. In some cases, this step may involve preheating dies to an operating temperature before installation.

This step can be simplified if the work area is properly organized and equipped. The preparations for each setup should be in a written, step-by-step procedure. Such organization allows all tools and equipment to be in the proper position before the change begins. Three simple but important reminders: 1) do not have to search for anything, 2) do not have to move anything, and 3) do not use anything that is not specified in the procedure. If these simple steps are followed, then the results should be substantial.

2.2 Internal Setups

Internal setups are performed when the equipment is not operating. This step is the actual changing or replacement of the toll, die, etc. The shorter the time needed for the internal setup, the faster the equipment can begin to produce a new product. In some cases, the amount of setup time has been reduced from days to hours and even to minutes. Manufacturers have developed a technique called SMED, which stands for single minute exchange of dies. Its goal is to reduce the internal setup time to as small as possible. The basic steps to improving internal setup time include standardizing work procedures, which enables the crews to be consistent in the change, with no unexpected problems.

Most setups require several people to be involved in the process. In order to insure the effectiveness of the setup, all work should be allocated to the people prior to the actual setup. They should be allowed to work in parallel paths to accomplish as much as possible in a short time period. Some companies will videotape this process and allow the employees to study the tapes, looking for ways they could improve the tools or techniques to reduce the actual setup time.

Additional areas to investigate include the physical equipment itself. Attempts should be made to reduce the number of parts required to clamp the equipment in place. Also the clamping method should be as simple as possible to allow quick removal and replace-

DEFINING EQUIPMENT LOSSES   41

ment, without compromising the necessary clamping force to produce a quality product. Considerations involve the shape, weight, and interchangeability of the equipment. If these areas are effectively analyzed, the setup time can be dramatically improved.

2.3 Adjustments

After the equipment has been changed, a period of adjustment is required to insure that the product meets the quality specifications. In most cases, this time can be considerable. The goal of studying the adjustments is to find methods that will eliminate all adjustments. If all factors are set to specifications during the internal and external setups, this goal can be achieved. One area of consideration is the precision of the equipment. Trying to get one more cycle out of a tool or fixture can be more expensive than the replacement would have been. The time lost trying to make marginal or out-of-tolerance equipment function properly will be considerable. In addition, all of these adjustments will have to be changed when the next setup is made because the good equipment will require all different settings.

One area where major improvement can be made at low cost is the simplification of the adjustment process. This step includes good reference surfaces, simple measurement methods, and simple standard adjustments. In many cases the go - no go gauges can be used to speed up the process.

If the equipment, tools, and procedures are kept simple and effective, the goal of no adjustments can be reached.

3.0 Idling and Minor Stoppages

Idling and minor stoppages occur when the equipment is stopped for a few minutes or is idle due to some upstream operation malfunction in a process or assembly line. Because these minor stoppages are too often ignored, they become accepted over a period of time as a normal operational characteristic of the equipment. The losses in this area can be considerable, in some case even more than the losses caused by the capacity loss breakdown. The three main types of idling and stoppages are:

1. Overloads
2. Equipment malfunctions
3. Upstream malfunctions

The overloading of equipment trips some type of limiting device. The equipment will then shut down until operations or maintenance personnel reset the device. This loss will only be for a few minutes, but, if repeated over time, can be greater than anyone realizes. For example, if a piece of equipment experiences a 10-minute stoppage per shift, on a 3-shift per day operation, running 5-days per week, the monthly total of downtime for the equipment is 10 hours. If you consider what an hour of downtime costs on a piece of equipment, the true cost of this type of loss can easily be in the thousands, tens of thousands, or even hundreds of thousands of dollars. These costs multiply even more rapidly if the equipment is not stand alone, but instead part of a production process. The delay on the first process will contribute to similar delays on the processes downstream. In a Just-In-Time (JIT) or Computer-Integrated-Manufacturing (CIM) environment, the total loss may be tremendous. Yet these small losses go unnoticed and uncorrected by most companies. In order to be cost effective, these losses must be eliminated.

The steps necessary to correct these minor stoppages are basic maintenance practices. Each delay should be investigated. Root cause analysis should be performed, with a fishbone diagram used to track the common problems. The common problems should be analyzed first, with the more complex problems being addressed after these are eliminated.

Proper basic maintenance will also help to eliminate these problems. In many cases, keeping the equipment in its proper repair and adjustment, thereby insuring that it is close to its optimum position in its life cycle, will help to eliminate most idling and minor stoppages. One key point should be remembered: "Act on each problem." Simple neglected problems can develop into serious complex problems.

4.0 Reduced Capacity Losses

Reduced capacity losses are defined as the difference between the rated capacity of the equipment and the capacity at which it is operating. The two largest areas that impact the capacity of the equipment are the speed and volume of the output.

The first problem occurs when the capacity of the equipment is unknown. This happens especially with older equipment. The true speed and output of the equipment should be identified and become the new capacity rating of the equipment. Then reaching the capacity will usually entail correcting many minor problems and defects that prevent the equipment from reaching this level. As with many other problems previously discussed, operations personnel learn all too often to accept less than design capacity from their equipment over long periods of time.

The second problem involves the failure of operations and maintenance to correct problems as they are uncovered. Allowing the equipment to limp to the next shutdown or outage, may cost more than if it is taken off line and the problem corrected. The lost production due to the decreased capacity, coupled with the increased costs for operations and maintenance, will be greater than any benefits achieved by continuing to operate the equipment at the reduced rate of production.

The rule to remember in this area is to not accept anything less than capacity-rated performance from any equipment. If this rule is followed, all necessary steps will be taken to correct minor problems before they become major problems.

5.0 Quality Problems

Maintenance-related quality problems can be divided into two broad classifications: habitual and occasional. Habitual defects are caused by equipment operating with some deterioration that has come to be accepted as normal operation. The occasional defect is caused by some apparent malfunction or problem that has suddenly developed.

Habitual defects are caused by equipment-related problems that have developed over a longer time period and are neglected. The defects may result from a series of short-term fixes or the "band-aid" approach to solving the problem. However, the true problem is never addressed. In order to compensate for the problem other adjustments are made. These adjustments affect other systems on the equipment and soon, the operational standards are affected. These defects will have varying effects on quality. Only by restoring the equipment to its original operating standards will the habitual defects and their quality-related problems be eliminated.

The occasional defect is cause by a failure in one of the systems or parts of the equipment. The problem and its cause are quickly and easily discerned. Corrective action to remedy the problem should be taken immediately. In no case should an occasional defect be neglected. The equipment should be restored at once.

In studies conducted at some plants, over 50% of all quality problems are maintenance related. Failure to correct both habitual and occasional equipment problems can lead to very expensive quality-related problems.

6.0 Startup/Restart Losses

These losses are incurred each time a process must be shutdown and restarted. They may involve producing an unacceptable product while the equipment reaches a certain operations parameter such as temperature or speed.

Startup and restart losses may be considerable; therefore, any equipment shutdowns related to a failure must be avoided. These losses are often overlooked when considering downtime costs. The costs are not just the lost production while the equipment was shut down. They also include the lost production while shutdown and startup were occurring. In some plants, these processes may involve hours of lost production per occurrence.

By applying many of the techniques already described, these losses can be kept to a minimum. In no instance should these losses fail to be included in the total lost production cost calculated during a function loss or function reduction breakdown.

7.0 Other Related Costs

There are other costs in addition to the six major areas of loss already identified. These losses include manpower loss and the manpower used at a premium cost to make up lost production and energy losses. These losses are self-explanatory.

In concluding this section, the combined costs related to the equipment and facility not being properly maintained can make up 30-50% of total production costs. From a facility standpoint, the loss can also be measured in poor worker productivity, poor appearance and image to clients, and lack of community support for the company.

Measuring the Equipment Performance

Measuring equipment effectiveness must go beyond just the availability or the uptime. It must factor in all issues related to equip-

DEFINING EQUIPMENT LOSSES 45

ment performance. The formula for equipment effectiveness must look at the availability, the rate of performance, and the quality rate. This approach allows all departments to be involved in determining equipment effectiveness. The formula could be expressed as:

Availability X Performance Rate X Quality Rate = Overall Equipment Effectiveness

Availability is the required availability minus the downtime, divided by the required availability. It can be expressed as the following formula:

$$\frac{\text{Required Availability - Downtime}}{\text{Required Availability}}$$

Required availability is the time production needs to operate the equipment minus miscellaneous planned downtime, such as breaks, schedule lapses, and meetings. Downtime is the actual time the equipment is down for repairs. It is sometimes called breakdown downtime. The calculation gives the true availability of the equipment, the number that should be used in the effectiveness formula. The goal for most companies is a number greater than 90%.

Performance rate is the ideal or design cycle time to produce the product multiplied by the output and then divided by the operating time, as seen in the following formula:

$$\frac{\text{Design cycle time X Output}}{\text{Operating time}}$$

Design cycle time (or production output) is given in some unit of production such as parts per hour. Output is the total output for the given time period. Operating time is the availability from the previous formula. The result will be a percentage of performance. This formula is useful for spotting capacity reduction breakdowns. The goal for most companies is a number greater than 95%.

Quality rate is the production input into the process or equipment minus the volume or number of quality defects divided by the production input. The formula is:

$$\frac{\text{Production input - Quality defects}}{\text{Production input}}$$

Production input is the unit of product being fed into the process or production cycle. Quality defects represents the amount of product that is below quality standards (not rejected, and there is a difference) after the process or production cycle is finished. The formula is useful for spotting production quality problems, even when the poor quality product is accepted by the customer. The goal for most companies is a number higher that 99%.

Combining the total for the three goals, we find that our goal for overall equipment effectiveness is:

$$90\% \times 95\% \times 99\% = 85\%$$

To be able to compete for the national TPM prize in Japan, overall equipment effectiveness must be greater than 85%. Unfor-tunately in most U.S. companies, OEE is barely more than 50%. It is little wonder that there is so much room for improvement in the typical overall equipment effectiveness analysis.

How senior management is informed about overall equipment effectiveness and learns to see value in the calculation will be covered in Chapter 3.

CHAPTER 3

OVERALL EQUIPMENT EFFECTIVENESS

Chapter two developed the formula for Overall Equipment Effectiveness:

OEE = Availability X Performance Efficiency X Quality Rate

The goals are for an availability of 90%, a performance efficiency of 95% and a quality rate of 99%. Multiplied together, these give an overall equipment effectiveness goal of 85%. When senior managers first hear a presentation about OEE, this number is "cold" in that they don't always understand the number correctly. To communicate OEE successfully to senior management, we must convert the OEE percentage into financial terms. This chapter presents case studies describing where this conversion was made, establishing a methodology you can use within your company.

Equipment Efficiency Case Studies

This case study looks at a disk drive manufacturing operation. During the process, a robotic arm picks up a computer hard drive and repositions it for transfer to the next operation. A problem developed where the robotic arm was dropping the disk drives periodically. The operators were becoming frustrated and wanted management to replace the robot. A team was assembled and given the task of improving the robot's operation. They learned that the robot was dropping a disk on the average of one every half hour of operation.

When the robot dropped the drive, it took an average of five minutes for the operator to reposition and re-index the robot. The operators were asked what would be an acceptable level of performance. After conferring, they replied that they could accept the robot dropping a drive once every two hours.

Let's examine their answer. A dropped drive once every two hours in a 24 hour a day/7 day a week operation adds up to a considerable amount of time. The operators worked a 12-hour shift. One dropped drive every two hours at five minutes downtime per drop adds up to thirty minutes downtime per shift. Multiplying this amount times the 14 shifts the robot operated equals 7 hours of downtime per week. In turn, 7 hours of downtime times $10,000 per hour of production output is $70,000 of downtime every week. Finally, multiplying this number times the 50 weeks a year the plant operates leads to an amount of $3,500,000 — and that is the amount the team said it was willing to tolerate! The problem with the robot was ultimately resolved and has reduced the downtime to almost zero.

This example highlights a problem. If there was one single occurrence of seven hours of downtime during a week of operation, everyone would focus on finding and fixing the problem. However, if you divide the seven hours of downtime into five-minute increments and distribute them throughout the week, people tend to ignore them, believing that this is the way the equipment is supposed to operate. Numerous studies have shown that these small losses add up over time, causing more of a loss than do the large breakdowns. Unfortunately most people ignore these small losses. However, using the overall equipment effectiveness calculation correctly can expose these small losses.

A second case about equipment efficiencies studied a water softener. A series of large water softeners were used to produce soft water that was turned into steam, then injected down a hole in an oil production facility to force out the crude oil for processing. In this study, the water softeners were the bottleneck, restricting the amount of steam that could be produced and, in turn, the amount of crude oil that could be extracted. Because the water softeners were the problem, management requested capital funding for the purchase of a sixth water softener to increase the volume. While the purchase was being processed, the field personnel put a cross-functional team together to examine the water softener problem. As they examined the OEE of the softeners, they discovered that the performance efficiency was very low (mid 50s). Examining the losses, the team found that the operators were rejuvenating the softeners on different frequencies and using different procedures to perform the rejuvenation. These procedures impacted the quantity and quality of the soft water the softeners could produce.

The team developed standard schedules and procedures, then trained the operators on these procedures. Immediately the volume of soft water increased to the point the water softeners were no longer the process bottleneck. The local management was able to return the capital appropriation for the purchase of the sixth softener because it was no longer required. The savings to the company totaled $750,000. As this example indicates, understanding the true design output of equipment can help avoid over-investing in assets. An important point to remember: expense dollars not expended become profit dollars.

Overall Equipment Effectiveness Case Studies

1. Plastic Injection Molding Press

This example is taken from an automotive parts supplier to one of the major U.S. auto manufacturers. The supplier ran a plastic injection molding press that produced components on a schedule of three 8-hour shifts, 5 days per week (see Figure 3-1). This schedule allows for a total of 7,200 minutes for possible production. The company had planned downtime of 600 minutes per week (20 minutes for lunch per shift plus two 10-minute breaks per shift), leaving a net available run time of 6,600 minutes per week. The total downtime losses averaged 4,422 minutes per week, leaving an actual operating time of 2,178 minutes. Availability is then calculated to be 33%. During the 2,178 minutes that the equipment ran, it produced 14,986 pieces. With a design cycle time of .109 minutes per piece, the operational efficiency is 75%. On average, there were 600 rejects during the week; therefore, the rate of product quality is 96%. When availability (33%) is multiplied times the efficiency rate (75%) and quality rate (96%), the overall equipment effectiveness is 24%.

Given that 85% is considered to be the goal for OEE, then the supplier had a large opportunity for improvement. How do you convince peers and executives, however, that 24% is not good and we need to increase OEE by another 61 percentage points? Convincing senior managers to make decisions with only percentage points for data is a daunting task. A better solution would be to present the improvement plan based on financial incentives. It is necessary, therefore, to work the problem over, inserting the numbers that would be necessary to achieve an 85% OEE. These calculations are shown in Figure 3-2.

The second example shows that in order to have 90% equipment availability, the downtime losses can not exceed 660 minutes. This improvement will increase the operating time to 90%. With the increased availability, the total output would increase to 51,770 pieces at a design cycle time of .109 minutes per piece. With these increased volumes, the quality rate increases to 99%, lowering the rejects to 518 pieces. The difference in production volumes between the 24% OEE and the 85% OEE is an increase of 36,784 pieces.

SCHEDULE FOR INJECTION MOLDING PRESS

1. Gross time available (8 x 60=480min.) 7200 min.
2. Planned downtime (for P.M., Lunch, Breaks) 600 min.
3. Net available run time (1-2) 6600 min.
4. Downtime Losses (breakdowns, setups, adjustments) 4422 min.
5. Actual operating time(3-4) 2178 min.
6. Equipment availability (5/3 x 100) 33%
7. Total output for operating time (pieces, tons) 14,986 pieces
8. Design cycle time .109 mins./piece
9. Operational effeciency (8 x 7 / 5 x 100) 75%
10. Rejects during week 600 pieces
11. Rate of product quality (7-10 / 7 x 100) 96%
12. OEE (6 x 9 x 11) 24%

FIGURE 3-1

REVISED SCHEDULE FOR INJECTION MOLDING PRESS

1. Gross time available (8 x 60=480min.) — 7200 min.
2. Planned downtime (for P.M., Lunch, Breaks) — 600 min.
3. Net available run time (1-2) — 6600 min.
4. Downtime Losses (breakdowns, setups, adjustments) — 660 min.
5. Actual operating time (3-4) — 5940 min.
6. Equipment availability (5/3 x 100) — 90%
7. Total output for operating time (pieces, tons) — 51,770 pieces
8. Design cycle time — .109 mins./piece
9. Operational effeciency (8 x 7 / 5 x 100) — 95%
10. Rejects during week — 518 pieces
11. Rate of product quality (7-10 / 7 x 100) — 99%
12. OEE (6 x 9 x 11) — 85%

FIGURE 3-2

This figure, although impressive, still needs to be taken a step further. Each piece has a selling price of $10.00. Multiplying the 36,784 pieces times $10.00 per piece yields a net difference in revenue of $367,840. This figure will get any executive's attention. Still there is one additional step: to annualize the revenue differential. It amounts to an improvement of more than $19 million annually. With this kind of a business case, the focus becomes what a company would spend to increase revenue by $19 million annually. $1 million? $5 million?

This particular company had ten presses with an almost identical OEE. Now the revenue differential grows tenfold. Further-more, the company was set to build a new facility next door to house additional presses needed because the current ones could not keep up with market demand.

Although this case study may seem unrealistic, it is true and representative of events that occur all too often in companies. Without a clear understanding of how OEE affects their asset base, many companies make poor decisions concerning their assets. Companies

52   TOTAL PRODUCTIVE MAINTENANCE

will continue to be competitive only if they clearly understand how their assets should be utilized to support their business goals and objectives.

2. Offshore Gas Compressors Case Study

This example is drawn from an oil and gas company. The compressors being studied were on a deep-water platform used to feed the gas output from the field into a pipeline to send it onshore. The problem with these particular compressors was that they were aging. Maintenance wanted to take them offline for a complete overhaul. Operations felt they could not afford the downtime and resulting lost production. The organization was at an impasse. After consulting with the area superintendent, it was decided to do a financial study. The cost of the overhaul and the resulting lost production would be weighed against the lost production caused by the decreased efficiency of the compressors.

Two compressors were on the platform. OEE results for each compressor are detailed in Figure 3-3 and Figure 3-4. The first com-

SCHEDULE FOR COMPRESSOR #1

1. Gross time available	168 hours
2. Planned downtime	0 hours
3. Net available run time (line 1-2)	168 hours
4. Downtime Losses (PM., breakdowns)	13 hours
5. Actual operating time (line 3-4)	155 hours
6. Equipment availability (5/3 x 100)	92%
7. Total output for operating time	45.5 MCFM
8. Design output	73.5 MCFM
9. Operational effeciency	62%
10. Rejected products	0
11. Rate of product quality	100%
12. OEE	57%

FIGURE 3-3

pressor's results (Figure 3-3) indicate that availability met the goal of greater than 90%. The operation efficiency, however, was only 62%, indicating lost efficiency due to internal wear of the compressor. Completing the calculation, we find that OEE was 57%

The quality rate was not a factor in this particular example. Because the meter that was used to determine volume shipped was on the outlet of the compressor, there was no opportunity for rejects. If the gas went into the pipeline, the money was in the cash register. This particular example, therefore, has a two-part OEE instead of the traditional three-part OEE calculation.

The second compressor (Figure 3-4) showed a lower availability, but had a higher operational efficiency. With quality not a factor again, OEE was 67%.

Now came the difficult part: establishing the business case. First, if the efficiency was improved, did they have the extra gas capacity to draw from in their well? The answer was yes because they had some wells restricted and others shut off. Therefore, additional capacity was available. Second, was the market sufficient to support

SCHEDULE FOR COMPRESSOR #2

1. Gross time available	168 hours
2. Planned downtime	0 hours
3. Net available run time (line 1-2)	168 hours
4. Downtime Losses (PM., breakdowns)	25 hours
5. Actual operating time (line 3-4)	143 hours
6. Equipment availability (5/3 x 100)	85%
7. Total output for operating time	105 MCFM
8. Design output	133 MCFM
9. Operational effeciency	79%
10. Rejected products	0
11. Rate of product quality	100%
12. OEE	67%

FIGURE 3-4

54 TOTAL PRODUCTIVE MAINTENANCE

the extra production? Again, the answer was yes because the company could sell whatever they could add to the pipeline without impacting the price of the gas.

The next step was to understand the gap between what the compressors were able to produce currently and what the output would be at 85% OEE. At the time, the weekly output from the two compressors was 150.5 MCFM, the output measurement of the gas. If the compressors were restored to 85% OEE level, the output would be 196.23 MCFM weekly, a difference of 45.73 MCFM per week.

This production volume then had to be converted into a financial term. At the time of the study, the gas price was $2,290.00 per MCFM, a potential difference of $104,721.70 per week. Annualized, this amounted to almost $5.5 million.

Next, the cost of the overhaul had to be determined. First, the maintenance labor and materials to perform the overhaul were determined. Next the lost production that would be incurred during the outage was calculated. The compressors would be overhauled one at a time to prevent the field from being shut down. The total es-

---

CRANK SHAFT LINE

1. Availability—Goal 90%

    (30 days X 24 hours x .90)   648 hours

2. Performance Efficiency

    A. (11,800 / Day or 491.67 / hour)   318,602 parts
    B. (10,000 / Day or 416 / hour)   269,568 parts
    C. (9,000 / Day or 375 / hour)   243,000 parts

3. Quality Rate —Goal 99% (99.99999%)

    A.=   315,416 parts
    B.=   266,872 parts
    C.=   240,570 parts

FIGURE 3-5

timated cost for the overhaul was $450,000.

Now, the return on investment can be estimated by comparing the cost of the overhaul, $450,000, to the projected additional throughput valued at approximately $5,445,500. In theory the payback period was approximately 30 days. Once these figures were presented to the superintendent, the decision was made to proceed with the overhaul. The superintendent had the finance department track the expenditures compared to the increased production after the overhaul. The payback was close to what was expected; the actual payback period was 28.1 days.

Consider the impact of converting the OEE to a dollar value. The maintenance department already tracked the efficiency of the compressors. The efficiency loss was already known prior to running the financial analysis. It was only when the loss was converted to dollars that a decision was made to move forward with the overhaul. Impasses between maintenance and operations can only be broken when the financial impact of the issue being debated is highlighted. Only then can the best decision for the company's profitability be derived.

3. Compressor Crankshaft Case Study

This case looks at a production line that makes crankshafts for a refrigerator compressor. The line was not able to make the target production rates, which were already lower than the design capacity of the line. The study revealed that the line was supposed to run 24 hours per day, 7 days per week. It was decided to look at this OEE on a monthly time frame, rather than by a shift, day, or week (See Figure 3-5).

Because OEE was monthly, a thirty-day window was examined. Thirty days times 24 hours per day produces 720 hours of availability. Availability should be 648 hours to achieve the goal of 90% availability.

Performance efficiency was not as easy to calculate. Maintenance, engineering, and production disagreed strongly about the actual design production rate of the line. Almost everyone at the plant believed the production rate was 375 per hour. Researching documentation at the plant, older production reports indicated the line had at one time operated at 416 parts per hour. After contacting the vendor and other companies that had the same line, it was determined that the design production rate was 491.67 parts per hour. In order to reduce the amount of disagreement about the design production rate, it was decided to calculate OEE in a best case/likely case/worst case scenario. Figure 3-5 shows how these calculations were made. The figures for both performance efficiency and quality rate show all three options.

The next step was to prepare the financial comparison. This step involved taking the production volumes for the past month and establishing the financial parameters. When the per-hour rates were projected for a 24-hour period and the best/likely/worst case production rates were compared, the results were calculated (see Figure 3-6).

When the difference between what could have been made was compared to what was actually produced, the production differences were calculated. Each part had a shelf value of $16 and the

---

DIFFERENCE FOR A 30 DAY TIME PERIOD COMPARED TO THE MAY ACTUALS (181,230)

A. 11,800 perday

$$315,181,230 = 134,185$$
$$\times \quad \$16.00$$
$$\$2146,960.00$$

B. 10,000 per day

$$266,872 - 181,230 = 85,642$$
$$\times \quad \$16.00$$
$$\$1,370,272.00$$

C. 9,000 per day

$$240,570 - 181,230 = 59,340$$
$$\times \quad \$16.00$$
$$949,440.00$$

FIGURE 3-6

REVENUE GENERATION ANNUAL RESULTS

A. $2,146,960.00 x 12 mo. = $25,763,520.00

B. $1,370,272.00 x 12 mo. = $16,443,264.00

C. $949,440.00 x 12 mo. = $11,393,280.00

FIGURE 3-7

production variances were multiplied times this amount to calculate the difference in dollars. The monthly variance ranged from a best case of $2,146,976 to a worst case of $949,440, as illustrated in Figure 3-6.

The final step in the calculation is to annualize the potential savings, as shown in Figure 3-7. The saving ranges from a best case of $25,763,712 to a worst case of $11,393,280. The question then asks what senior management would be willing to invest to see the increases in production and the subsequent financial return.

In this case study, the final summary was presented to senior management. The numbers were not all that surprising to them. In fact, the Chief Financial Officer said he know of lines in other companies that ran at the best case production rates and thought the numbers that were presented were believable.

This company essentially had no formalized maintenance business process in place. Based on the return on investment presented, the senior management team decided to fund the improvement necessary to achieve the levels of production that were possible.

Summary

This chapter has shown the true value of converting OEE into financial terms for communicating with senior management. If this step is not taken, very few TPM programs will have sustainable results. The major reason for this is the revolving door policy in most managerial positions today. For example, a company may have a senior management team in place that clearly understands the benefits of TPM. However, this team is only in place for a short time. Because they produce results at the plant they are currently managing, they will be promoted. When the next management team comes in, they will want to do things their way. Unless they can be shown the benefits derived from TPM and the financial benefits that have been achieved and sustained, they will begin to disassemble the TPM initiatives. This is not conjecture – it has happened numerous times to companies that at one time had award-winning TPM programs in place.

The techniques in this chapter, when applied, should help produce TPM efforts that provide tremendous return on investment for the company and sustainable results far into the future.

CHAPTER 4

ACTIVITIES BEFORE TPM

Several activities must take place prior to beginning a TPM program. Most of these activities are designed to give the organization the information structure and discipline needed to successfully implement TPM. They include:
- Developing the maintenance organization
- Equipment information management
- Inventory information management
- Organizational assessment

Developing the Maintenance Organization

Developing the maintenance organization is an important step before beginning a TPM program. However, it is a step that many organizations, caught up in the spirit of starting a new program, often neglect. Because most of the equipment information and operator training comes from the maintenance organization, it is only logical to insure that it is properly functioning before turning attention to the operations and production departments.

Developing the maintenance organization is important in four main areas:
    a. Work order systems
    b. Information management
    c. Organizational structure
    d. Staffing

Proper development first focuses on the goals and objectives for the maintenance department. You may know where you want to go, but without knowing where you are, it is difficult to start. The objectives for the maintenance department may be basic, becoming more advanced as the organization matures. For example, the following objectives may be typical for the maintenance department:

    a. Attain the maximum production (or condition of the facity) at the lowest cost, while producing the highest quality and at the optimum safety standards.
    b. Identify and implement cost reductions
    c. Provide accurate equipment/facility maintenance records
    d. Gather and record all necessary maintenance cost information
    e. Optimize all maintenance resources

Objectives will vary from plant to plant or facility to facility, but clear objectives must be established for each maintenance organization. The objectives above should be used only to help each organization focus on its own objectives. It is difficult to state globally what the objectives must be. As a support function, maintenance must tailor its objectives to support and complement the goals and objectives of the rest of the corporation.

MAINTENANCE MANAGEMENT CONTROL SYSTEM

- Establish Goals, Objectives, Policies, and Procedures
- Establish Permissible Variance from the Guidelines
- Measure the Performance and Compare to the Guidelines
- Compare the Evaluation to the Permissible Variance
- Identify the Exceptions to Tolerance
- Determine the Cause for the Exception
- Determine the Corrective Action
- Plan the Implementation of the Corrective Action
- Schedule the Implementation of the Corrective Action
- Implement the Corrective Action
- Evaluate to Avoid Overcompensation or Failure of Action

Figure 4-1 A Sample maintenance management control System.

Once the objectives have been established, it is necessary to set a control system in place to monitor and report on the progress made in achieving these objectives. A typical control system is listed in Figure 4-1. This control system is designed to find exceptions to the stated objectives and supporting strategic programs. For example, a supporting strategy to identify and implement cost reductions might be the control of maintenance overtime. A good goal is 5% of total time worked. If the control system is used, no action is needed when the overtime is 5% or less. If the overtime grows over 5%, then an exception report should be generated showing that one of the indicators has been exceeded. The true problem should be discovered and a corrective action implemented. The overtime indicator should then be monitored to be sure the corrective action actually solved the problem.

The challenge in this case is that maintenance problems typically have multiple causes. Diagnosing them is similar to doctors diagnosing a disease. Many diseases have similar or identical symptoms. Only by running various tests and using different techniques to actually determine the problem can doctors make a proper diagnosis. They can then treat the true disease quickly and effectively.

In maintenance organizations, the same problem exists. Instead of solving the true problem, the symptom is treated. For example, if maintenance costs increase, the first action is to reduce headcount or inventory levels. It may be that these will appear to solve the prob-

---

COMMON MAINTENANCE EFFICIENCY PROBLEMS

- Equating Maintenance and Production Staffing Levels
- Crew Location Problems: Area—Central—Combination
- The Effect Reactive Maintenance Has on Productivity
- High Levels of Unexpected Downtime and the High Cost
- Random and Seasonal Fluctuations
- Poor Communications: Maintenance—Operations—Management
- Lack of Equipment Standards
- Lack of Qualified Personnel

Figure 4-2 Common maintenance efficiency problems.

lem, but the disease still exists. Once the organization stabilizes after the reduction, the disease will surface again. It is only by using techniques requiring multiple indicators that the true problem can be uncovered. Consider the typical problems listed in Figure 4-2. How many of these problems are related? How many have the same symptoms? Often two or more problems are interrelated. Only by having good, reliable data will the true problem ever be uncovered and solved.

How does the organization decide on what information to track and then gather that information? It starts by understanding the corporate goals and objectives, then structuring the maintenance goals and objectives to support them. Once the maintenance goals and objectives are established, the data required to monitor the results is identified, usually in the form of a maintenance performance indicator. In turn, the data necessary to support the indicator is identified and then it is determined how to collect the data. Some data will come from operations or facilities or even financial. However, the majority of the data necessary to support the indicators must come from the maintenance department itself.

Highlighting this point is the typical problem discovered by the financial officers, "Maintenance costs are up this month." The next question is why? The financial department may say we used more materials than normal. The next question is why? The financial department may be able to determine the cost center or account code from which the parts were used, but only the maintenance department knows what job or piece of equipment on which they were used. The maintenance department tracks this information through the use of a comprehensive and disciplined work order system.

Work Order Systems

The maintenance work order system is analogous to the production order system. In the latter, the order number is used to track a customer's request from the initiation in sales through the entire plant and out to the customer's site through shipping. The production order is filed so that any questions or problems with the customer's orders can be resolved quickly and correctly. It is also valuable for when the customer places another order for the exact same items: the order is pulled from history and duplicated.

ACTIVITIVES BEFORE TPM   63

In the maintenance work order system, the customer (operations or facilities) places the order for a repair or service. The maintenance department assigns the work an order number. Then all activities related to the completion of the work order are recorded on the work order. When the work order is completed, all costs are posted to it and it is filed in history. If the request is ever placed again, then the history can be reviewed and the order completed as before.

Just as reports and analyses are conducted on the production system to ensure the profitability of the company, the same is done to the maintenance information. This ensures that complete, cost-effective results have been achieved. Any problems or questions are easily referenced and resolved to the customer's satisfaction.

This scenario should help everyone to appreciate that maintenance not only can he, but needs to he, managed and controlled like other parts of any business. The work order is the information-gathering device used to feed data to the maintenance control system. The typical objectives for work order systems are listed in Figure 4-3.

While the information necessary to accomplish these objectives may vary from site to plant to facility, the information gathered on the work order should always be focused on supporting the information needs. Miscellaneous information that does not contribute to satisfying the needs only complicates the information-gathering process. In most cases, too much information is just as frustrating as too little information. This is why the information needs and the control system indicators must be identified early in the process.

WORK ORDER OBJECTIVES

- A Method for Requesting, Assigning, and Following Up Work
- A Method of Transmitting Job Instructions
- A Method for Estimating and Accumulating Maintenance Costs
- A Method for Collecting the Data Necessary for Producing Management Reports

---

Figure 4-3 Examples of objectives for work order programs.

Information Management

After the control system is set in place and the work order system is producing information, the next step is to manage the information. In the early days of computerized maintenance management systems; companies gathered a lot of data, but no one knew what to do with them. All the information gathered should have a purpose. If the indicators are identified during implementation of a program, the tendency to gather too much data is eliminated.

The information gathered should be identified as to the timeliness of the reports that will be created. The typical distribution for reports is on a daily, weekly, monthly, quarterly, and annual basis. When creating maintenance reporting requirements, the previously mentioned point must be reinforced. Exception reports are used to manage, summary reports are used to review trends, and detailed reports are used to archive data and review specifics.

Each of these reports has its use. But under no circumstance should managers have all types of reports on their desk each day. Being buried in reports—not having the time to understand the data provided—is just as bad as not being given any data at all. In either case, the manager will not have the complete picture.

One additional point about the reports regards the accuracy of the data. These reports will only benefit the company if a disciplined approach is taken to the work order system. If repairs are reported as preventive maintenance tasks, if parts are charged to the wrong work orders, and if emergencies are reported as normal repairs, then the true picture will always be distorted. The analysis is only as good as the data. Dedication to the basics—such as gathering accurate information—will ultimately spell success or failure of any maintenance improvement program.*

Organizational Structure

The structure of the maintenance organization is important to the effectiveness of the service it provides. Three common types of organizational structures are centralized, area, and combination. Each type of structure has its own advantages and uses.

---

*For a complete description of reports that may be used to manage maintenance, see the book Benchmarking Best Practices in Maintenance Management (New York: Industrial Press, 2003).

Centralized Organizations

The centralized organization is generally found in companies that are built around a facility or plant that is confined geographically. Maintenance workers have very little travel and can reach all equipment in a relatively short period of time. This arrangement is usually supplemented by a centralized maintenance stores arrangement, allowing for lower maintenance inventories and good service from the stores counter.

Area Organizations

Area organizations are used where the equipment is spread out over a larger area and travel time becomes a problem for the maintenance personnel. Communication and logistics problems increase in an area organization. However, area organizations tend to build team spirit and cooperation between the area maintenance workers and the operations personnel. This step is important to the beginning of a TPM program. The maintenance workers also begin to develop ownership in the equipment—another plus in the TPM team development. A problem that may develop in this organization is the concept of area stores. These stores increase inventory levels and require additional storage areas. However, as maintenance becomes more proactive, and as planning and scheduling become more disciplined, a delivery system can be set up that will reduce the need for too many area stores.

Combination Organization

The combination organization allows the flexibility of the centralized organization and still gains the benefit of the area organization. The team work and ownership of the area organization are still fostered, but the support from the central organization is available when needed. This organizational arrangement is the last step maintenance takes before entering a team-based organization. The maintenance team member coordinates all maintenance activities for the team, utilizing the centralized manpower as required to complement area efforts.

The combination organization is the goal that each larger company should work toward in order to set the foundation for TPM. The smaller company may still stay centralized, but could assign the preventive maintenance and repair responsibility to certain individuals. In one respect, these become "quasi-area" assignments. This arrangement still enables the ownership and teamwork concepts to begin to develop.

Staffing

The staffing of the maintenance organization should be divided into two separate groups: craft technicians and support personnel.

Craft Technicians

The craft technician staffing is determined strictly by the backlog of work for each craft. In a multi-skilled environment, the backlog may be the total maintenance workload. The craft backlog is the amount of work that is documented as needing to be performed by the craft. The work should be documented on the work order form; the work that is counted in the backlog is only the work that is ready to schedule or that can be performed at the present time. This removes work that is waiting on engineering support, spare parts, approval, shutdowns, outages, or rebuilds.

The formula for calculating the backlog is as follows:

$$\text{Backlog (in weeks)} = \frac{\text{Total planned hours ready to schedule}}{\text{True craft capacity}}$$

The true craft capacity is the total hours scheduled for the craft for a week minus schedule interruptions. Schedule interruptions should include average hours spent on emergencies, absenteeism, vacations, and routine or (in some cases) preventive maintenance work. This leaves the total hours that the craft will actually deduct from the backlog. An example of this process is illustrated below.

Total employee hours scheduled for next week (10 techs x 40 hours)	400 hours
Total overtime to be worked (average for last 3 months)	40 hours
Total contractor labor (2 techs x 40 hours)	80 hours
Gross labor hours available	520 hours
Average emergency work (50% for the last 3 months)	260 hours
Average absenteeism/week	10 hours
Average vacation hours/week	10 hours
Average routine (nonbacklog) hours/week	40 hours
Total deductions	320 hours
Gross minus deductions = 200 hours	

These 200 hours represent what can realistically be expected to be completed from backlog work for the week. This is the number that should be used to determine the true backlog. The calculation would be as follows (assume 2000 hours in the backlog):

	Example 1	Example 2
Backlog in weeks =	2000 hours / 520 hours	2000 hours / 200 hours

In Example 1, the craft backlog is 3.8 weeks. But it is impossible to complete the work in the backlog given the constraints placed on the craft time available.

In Example 2, the craft backlog is 10 weeks, a realistic time period in which to complete the work. Normal maintenance backlogs are optimized when they are kept in the 2-4 week range. Some companies will allow a 2-8 week backlog for a craft. The greater the number of weeks in the backlog, the longer the requesting department has to wait on its work request. If the backlog gets too large, the temptation is to request the work on an emergency basis, circumventing the planning and scheduling process, and thereby increasing the cost to do the work.

If the backlogs are kept in the 2-4 week range, the maintenance disciplines are easier to maintain. If backlogs begin to increase, then overtime and contract labor may be used to see if the increase is temporary. If the increase is a permanent trend, new craft technicians may have to be hired. If the backlog decreases within two weeks, the overtime should be eliminated, and perhaps contract labor decreased or eliminated. If the backlog continues to decline, then the staffing may need to be examined and employees repositioned. Under no circumstances should maintenance labor be increased or decreased without accurate backlog figures. Changes made independently of the backlog are arbitrary and dangerous to the effectiveness of the maintenance organization and ultimately to the corporation.

Staff Personnel

The proper level for staff support is one of the most debated topics in maintenance. The correct method of determining staff levels begins with the correct determination of the craft staffing. Once the total number of craft technicians is determined, the correct level of first line supervisors and planners can be determined. The normal ratios are 1 planner for every 15-20 craft technicians, and 1 supervisor for every 10 craft technicians. Therefore, an organization of 40 craft technicians would require:

4 supervisors
2 planners
1 manager

The number of clerks is determined by the amount of data the organization is required to keep to meet its information management objectives. Care must be exercised not to overstaff in this area. In some cases, the number of clerks increases because the planners treat them as their secretaries. Planners will handle most of their own paperwork. Clerks are supposed to help manage the information flow, such as reporting, filing, and timekeeping. If everyone is doing their job in the example above, the clerical load may be one, or at the most two.

As a rule of thumb, staff levels for a maintenance organization should never exceed 25% of the craft workforce (assuming that the number of craft technicians is correct). The level of clerical staff support is ultimately determined by the amount of data the organization is required to keep. If a company requires maintenance to perform asset tracking, then the clerical function will be staffed at a higher level because the data tracking requirements are much higher. Geographic layout of the plant, skill levels, sophistication of the equipment, and other factors prevent any firm rules for staffing to be specified. Each plant, even within the same company, should be studied to determine the correct staff levels.

Equipment Information Management

Because a TPM program is designed to increase equipment effectiveness, it is important to identify the particular asset or equipment that is to be maintained. The activities for equipment information management that should take place before TPM are:

    a. equipment identification
    b. nameplate information
    c. equipment history

Equipment Identification

Equipment identification starts with deciding the difference between a piece of equipment and a component or part. In some companies, a motor is a piece of equipment; in other companies, it is a part or component. Why is this distinction a problem? Each piece of equipment will be required to have a detailed history, made up of the completed work orders for the piece of equipment. While the parts or components will require some data to be kept, the volume of information required for parts and components is less than that required for equipment. An equipment item should require a detailed historical repair record. The repair cost, frequency, date of repair, year-to-date costs, and life-to-date costs are all common data that are kept. The more equipment that is listed, the greater the clerical load. Some companies set a dollar limit or size limit on various components in order for them to be classified as equipment.

Once the items have been classified as equipment, they must be given a specific equipment identifier, usually called the equipment number. It is important to use an intelligent equipment numbering scheme; otherwise, it will be difficult later in the program to track the information required to improve equipment effectiveness.

Sequential numbering schemes are commonly used to identify the equipment. Although these may seem to work for just tracking data, they fail to provide the flexibility that is required for reporting. The following example provides details on numbering schemes and their usefulness.

When numbering equipment, it is wise to consider the layout of the plant or facility. The first set of digits could be used to indicate the area (see Figure 4-4A). The second series of digits could be used to identify a specific unit within that area (see Figure 4-4B). The third series of digits could be used to indicate the type of equipment to which the number applies (see Figure 4-5A). The last set of digits would signify the particular equipment item that is being identified (see Figure 4-5B). For example, when the whole series is put together, the number would be:

006-001-001-003

This sequence represents the number 3 hydraulic system on unit 1 in area 6. While this seems like a lot of work, there is a reason

70 TOTAL PRODUCTIVE MAINTENANCE

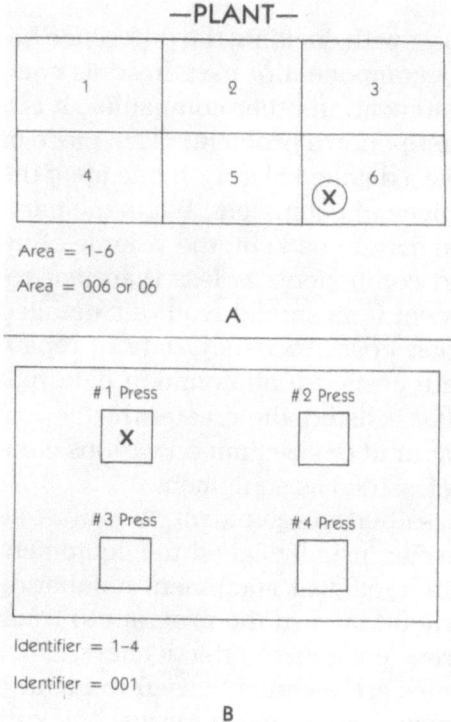

Figure 4-4 Areas of a facility and specific equipment in an area can be assigned individual number codes for easy identification

for it all. In a manual system, the numbers will not be useful other than to identify the specific equipment. But in a computerized system, it allows tremendous flexibility for reporting. For example, if you wanted to find all the breakdowns on hydraulic systems in the plant, you know that the third series of numbers is the equipment type, and the identifier for hydraulic systems is 001. You can write a simple query for the computer to find all repair records for type 001. This flexibility is lost when sequential numbering schemes are used.

A warning about numbering schemes: keep them simple. If they are too long (over 12-15 digits) or are unintelligent, they will be difficult to remember and will quickly fall into disuse. The best numbering schemes are short and intelligent.

Nameplate Information

After the equipment has been identified, it is necessary to gather information about each individual equipment item. This information

ACTIVITIVES BEFORE TPM   71

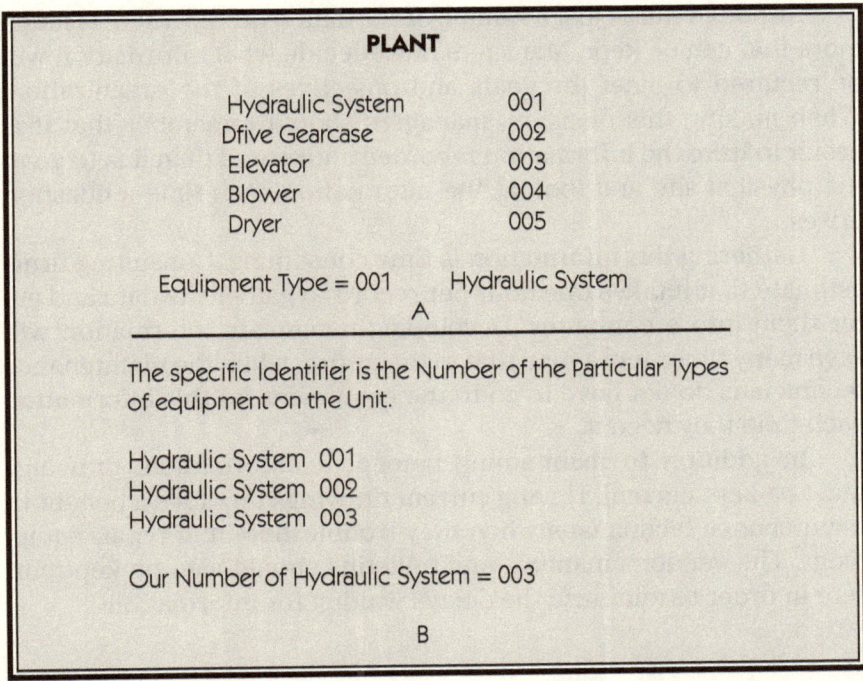

Figure 4-5 Additional codes can be used to identify the type of equipment and particular functions.

focuses on the components that make up the equipment. Continuing with the example of the hydraulic system, the following list provides some of the nameplate information:

Motor ID #	2884-3848
RPM	1800
Frame	32-T
Voltage	440 VACS
Shaft OD	1~/s"
Bearing Type	Fafnir 803
Pump ID #	234-341
RPM	1800
GPM	200
Frame	486-D
Manufacturer	Vickers

Although this is just a sample of the data required, there is much more that can be kept. Managers must decide what information will be required to meet the goals and objectives of the organization. When making this decision, managers should remember that it is easier to have the information recorded and stored than it is to go to the physical site and look at the information each time a question arises.

Gathering this information is time consuming. Consulting firms estimate that it takes one hour per record to gather the data and enter them into a computer. Developing nameplate information will save many times more than that in the future, when the maintenance technicians do not have to go to the equipment for this information each time they need it.

In addition to maintaining nameplate information, drawings must be kept current. Having current drawings provides a benefit to maintenance technicians when they troubleshoot and repair equipment. The vendor's manuals and bulletins should also be kept current in order to minimize the delays waiting for information.

Equipment History

The equipment history is a log of repairs kept for each equipment item. It enables repair records to be used to help determine preventive maintenance frequencies, overhaul intervals, and replacement cycles. Some of the information that needs to be kept for the equipment history includes:

> Type of work
> Crafts involved
> Date of the work
> Dollar value of the repair
> Any specific detail information

This information needs to be compiled into a useful format that can be quickly analyzed by the managers, supervisors, or craft technicians. Computers are a big aid for processing the data. Summary reports and exception reports are most useful for finding equipment that needs analysis or repair.

ACTIVITIVES BEFORE TPM   73

The equipment history is the true goal for the work order system. It is the database from which most decisions will be made. The validity of the decisions will depend completely on the accuracy of the data that are gathered. If work order history is available at the start of TPM, the accuracy of the information is vital. If the information is not reliable or accurate, it should not be added to the history. Subsequent parts of the TPM program will be based on this information, and either inaccurate or unreliable information will be damaging.

Inventory Information Management

Inventory information management is similar to equipment information management, except that its focus is on gaining control of the information about maintenance spares and inventory items. It uses essentially the same three basic steps as equipment information management, as follows:

   a. Part identification
   b. Part information
   c. Part history

Part Identification

This step is the process of numbering all the inventory items with a specific scheme for identification. Some companies will use the vendor's or manufacturer's part number or part identification. This approach is not a good one. From time to time, companies will change vendors, manufacturers will change part numbers, or items will be discontinued. Because the part number is a key number used for tracking all related information, it must never he changed. Particularly with computerized inventory systems, in which the part number is a record key, changing the number results in loss of history, detail information, and equipment cross references. Each company should develop its own part numbering schemes. A sample might look like:

030-   005-   0235
(A)   (B)   (C)

The A grouping is the particular classification of the item, for example electrical, electronic, power transmission, hydraulic, or pneumatic. The B grouping is the type within each classification. For example, the power transmission group might have types such as roller chain, V-belts, flat belts, and gear drives. The C grouping is the specific part.

The example above could be a V-belt, translated as follows:

    030 = power transmission
    005 = V-belts
    0235 = specific belt 4L-235

Each company must evaluate the numbering scheme it wants to develop; again, as with the equipment, it should be an intelligent numbering scheme.

Part Information

The information required to be kept for each part includes the following:

    Locations
    Quantities
    Details
    Manufacturers
    Vendors

This information is very similar to the detailed nameplate information that is kept for the equipment. One major difference is specifying the locations where the parts are stored. These locations should be specific so that when the parts are needed, they can be quickly and easily found. If an inventory storage area is unorganized, finding the part and recording the detail information can take approximately one hour per item.

Part History

The part history is the usage of the part for the last 12-24 months. The usage patterns that are recorded can help set the ordering patterns, stocking levels, and purchasing quantities. This information can be gathered from purchasing if it is not presently kept by the stores group.

The part history should not be used unless it is accurate. If the information is unreliable, then all the decisions involving parts will be unreliable.

Organizational Assessments

By the time a company has gathered the above-mentioned information, it will have detailed information about its plant's or facility's present condition. But the information is still fragmented. The next step is to perform an organizational assessment to pull the information together. The organizational assessment used to be called a maintenance audit, but that name is really a misnomer. Because maintenance is a service organization, many of its attitudes, ideas, and methodologies are influenced by other parts of the organization. Any influencing factors must be taken into consideration while assessing the condition of the maintenance organization. It is truly an examination of the attitudes of the company toward maintenance. In order to assist companies in assessing their maintenance attitudes, the maintenance management maturity grid was developed. This grid, patterned after the "Quality Management Maturity Grid" published by Phillip Crosby in the book Quality is Free, is pictured in Figure 4-6.

In addition to the grid, some of the following indicators on page 78 should be used to help benchmark the maintenance organization. Once these base indicators, or the utilization of other indicators, have been established, the plan to develop the maintenance organization can be formulated. The present benchmark should be documented and the projected improvements established.

Having the facts and projections makes the next step—the presentation of the plan—easier to prepare. Developing the plan to implement TPM requires understanding where you are starting. This is one of the most important steps before beginning the TPM program.

76 TOTAL PRODUCTIVE MAINTENANCE

The Maintenance Organizational Maturity Grid

Measurement Category	Stage 1 Uncertainty	Stage 2 Awakening	Stage 3 Enlightenment	Stage 4 Wisdom	Stage 5 Certainty
Corporate/Plant Management Attitude	No comprehension of maintenance prevention; fix it when it's broken [1]	Recognizes that maintenance could be improved, but is unwilling to fund [2]	Learns more about R.O.I.; becomes more interested and supportive [3]	Participative attitude; recognizes management support is mandatory [4]	Includes maintenance as a part of the total company system [5]
Maintenance Organization Status	REACTIVE: Works on equipment when it fails; otherwise very little productivity [1]	CONSCIOUS: Still reactive but rebuilds major components and has spares available when failures occur [2]	PREVENTIVE: Uses routine inspections, lubrication, adjustments, and minor service to improve equipment M.T.B.F. [3]	PREDICTIVE: Utilizes techniques such as vibration analysis, thermography, spectrography, N.D.T., sonics, etc., to monitor equipment condition, allowing for proactive replacement and problem solving instead of failures [4]	PRODUCTIVE: Combines prior techniques with operator involvement to free maintenance technicians to concentrate on repair data analysis and major maintenance activities [5]
Percentage (%) of Maintenance Resources Wasted	30+% [1]	20–30% [2]	10–20% [3]	5–10% [4]	Less than 5% [5]
Maintenance Problem Solving	Problems fought as they are discovered [1]	Short-range fixes are provided; elementary failure analysis begins [2]	Problems solved by input from maintenance, operations, and engineering [3]	Problems are anticipated; strong team problem-solving disciplines are utilized [4]	Problems are prevented [5]

ACTIVITIVES BEFORE TPM   77

	[1]	[2]	[3]	[4]	[5]
Maintenance Workers, Qualification and Training	Poor work quality accepted; rigid craft lines; skills outdated, skills training viewed as unnecessary expense; time in grade pay; low worker turnover/apathy	Workers' lack of skills linked to breakdowns; trade/craft lines questioned; skills obsolescence identified; training needs recognized; traditional pay questioned	Quality + Quality = Quality; expanded/shared job roles; a few "critical skills" developed; training expenses reimbursed; new pay level for targeted skills; increased turnover/fear of change	Quality work expected; "multiskill" job roles; skills up to date and tracked; training required and provided; pay for competency progression	Pride and professionalism permeate; work assignment flexibility; skilled for future needs: operators trained by maintenance, ongoing training; percent of pay based on plant productivity; low employee turnover/high enthusiasm
Maintenance Information and Improvement Actions	Maintenance tries to keep records, disciplines are not enforced, poor data	A manual or computerized work order system is used by maintenance, little or no planning and scheduling	A manual or computerized work order system is used by maintenance, operations, engineering; planners used; scheduling enforced	A computerized maintenance control system is used by all parts of the company; information is reliable and accurate	A maintenance information system is integrated into the corporate operation
Summation of Company Maintenance Position	"We don't know why the equipment breaks down; that is what we pay maintenance for. Sure, our scrap rates are high, but that's not a maintenance problem."	"Do our competitors have these kinds of problems with their equipment? Scrap is costing us a bundle!"	"With the new commitment from management, we can begin to identify and solve problems."	"Everyone is committed to quality maintenance as a routine part of our operational philosophy. We can't make quality products with poorly maintained equipment."	"We don't expect breakdowns and are surprised when they occur; maintenance contributes to the bottom line!"

Figure 4-6 The maintenance organization maturity grid.

TOTAL PRODUCTIVE MAINTENANCE

1. Supervisor to Technician Ratio	1 to 10-15
2. Planner to Technician Ratio	1 to 15-20
3. Labor Productivity	> 60%
4. Planning Efficiency Labor	> 90%
Materials	> 90%
5. Scheduling Efficiency (Weekly)	
Labor	> 90%
Materials	> 90%
6. Percentage of Overtime	< 5%
7. Size of Craft Backlog	2 to 4 weeks
8. Service Level of Stores	95-97%
9. Equipment Availability	> 95%
10. Percentage of Maintenance Activities Covered by a Work Order	100%
11. Percentage of Planned Hours Compared to Total Hours	> 90%
12. PM Costs Compared to Total Maintenance Costs	> 30%
13. Labor to Material Cost Ratios	50% / 50%
14. Maintenance labor Costs Compared to Estimated Replacement Value of the Plant	1%
15. Maintenance Materials Cost Compared to Estimated Replacement Value of the Plant	1%
16. Total Maintenance Costs Compared to the Cost of Goods	Varies
17. Maintenance Cost per Unit Produced	Varies
18. Maintenance Cost per Square Foot Maintained	Varies
19. Production Loss Caused by Maintenance	Varies
20. Union Attitude	Varies
21. Management Support	Varies
22. Corporate Environment	Varies

PART 2

IMPLEMENTING TOTAL PRODUCTIVE MAINTENANCE

In this portion of the book, we will examine the process necessary for a company to implement Total Productive Maintenance (TPM). This process will include a clear understanding of the maintenance basics that must be in place and how these activities must support TPM. We will examine the solutions to many of the typical problems encountered, from the maintenance perspective, while implementing TPM. This section will conclude with the typical problems encountered while implementing TPM and optional solutions to the problems.

CHAPTER 5

DEVELOPING THE TPM IMPLEMENTATION PLAN

Thus far, this book has discussed the theories and philosophies of Total Productive Maintenance as well as the benchmarking basic maintenance indicators. The vision of TPM for a particular site must start with these steps because 1) it is hard to develop a TPM plan if you do not understand what it is and 2) you cannot set TPM as a goal unless you know how far you have to go to reach it.

The following chapters cover the steps needed for implementing TPM. Each company can use these steps to design its own implementation plan. Most steps can be modified or altered to meet the needs of each site. All can be rearranged, if necessary, to suit the priority of each company. Chapter 5, however, is the exception. Gaining management understanding and support, then keeping it, is the highest priority for any organization. If all the reasons for the failure of maintenance improvement programs, including TPM, were compared, lack of management understanding and support would lead the list by far.

The ten-step plan presented in Figure 5-1 is designed to work with international companies. The approach is different than many traditional implementation approaches. The management style of international companies, their reporting requirements, the financial restraints, and the quarterly profit reporting all require a different adaptation of techniques and technology. The rest of this book provides a detailed explanation of these steps.

TEN-STEPS T.P.M. PLAN

- Developing the Long-Range Plan
- Selling the Program
- Ensuring Equipment Reliability
- Maintenance Inventory Controls
- Improving Maintenance Efficiency
- Maintenance Automation
- Training for C.A.T.'s
- Optimizing Maintenance Resources
- Team-Based Maintenance
- Measuring the Results

Figure 5-1 The ten-step plan for implementing a total preventive maintenance program.

Developing the Long Range Plan

Developing the long range plan requires looking at the corporate strategic plan and then developing a maintenance plan to support it. Once the analysis of the organization has been completed (see Chapter 4), the benchmark is set for plan development. By carefully analyzing the optimum ranges for some of the indicators, areas requiring improvement can be identified. However, the primary consideration should be the overall equipment effectiveness formulas. All of the maintenance indicators should be used to help find problems when the overall equipment effectiveness is low in particular areas.

Achieving the highest equipment effectiveness possible is the primary goal for all organizations within a company, including maintenance. Therefore, all of the indicators used to measure performance should ultimately be related back to the equipment. For example, most companies are concerned with labor productivity. How are OEE and labor productivity related? Low labor productivity indicates a lack of planning and scheduling, usually due to a high level

DEVELOPING THE TPM IMPLEMENTATION PLAN

of emergency work. High levels of emergency work will be due to a large amount of breakdowns (either capacity loss or capacity reduction). The equipment availability (uptime) is one of the three main components of the equipment effectiveness formula. Therefore, the indicator (labor productivity) will help identify the problem area.

The selection of the indicators that will be used to measure maintenance performance should be related back to the equipment effectiveness. This prevents the selection of arbitrary indicators over which maintenance has no control. In determining the long range plan, consider the following example.

AREA OF CONSIDERATION	PRESENT	GOAL
Organization	Centralized	Area - Team
Supervisors	1:5	1:10
Planners	None	1:20
Training Dollars	$200/Employee	$1000/Employee
Training Dollars	$500/Supervisor	$1200/Supervisor
Work Order Utilization	50%	100%
Equipment Effectiveness	45%	90%

The table above summarizes the indicators and goals for the long range plan. The maintenance indicators selected must be tied to overall equipment effectiveness.

Translating the Long Range Plan

The next step in developing the plan is to translate it into business terms and, more specifically, financial terms. By themselves, terms such as labor productivity, availability, and equipment effectiveness can be vague, misunderstood by senior management. Translating these maintenance terms into business terms involves the preparation of a cost justification. The following section provides an example for those unfamiliar with the process.

SYSTEM COST JUSTIFICATION

Part #1 EQUIPMENT DOWNTIME COSTS

1. Annual sales per year (Total) _____

2. Annual production labor costs per year _____

84 TOTAL PRODUCTIVE MAINTENANCE

3. Potential savings due to lost sales and production costs per year (1 + 2)   _____

4. Percentage of maintenance downtime per year
(Try to figure capacity loss and capacity reduction breakdowns)   _____

5. Potential savings from downtime reduction (3 X 4)   _____

6. Percent of improvement possible through improved maintenance controls
   - 0% - 10% if good maintenance practices are already in place
   - 10 - 20% if a basic manual system is already in place
   - 20 - 30%+ if a weak or informal system is in plac   _____

7. TOTAL DOWNTIME COST SAVINGS   (5 X 6)   _____

Part #2 LABOR SAVINGS

Percentages:
1. Time wasted by personnel looking for spare equipment parts
[use table below if actual is not known]
   - 15-25% No inventory system
   - 10-20% Manual inventory system
   - 5-15%   Work order system and inventory system
   - 0-5%   Computerized inventory and manual work order system   _____

2. Time spent looking for information about a work order
   - 10-20% No work order system
   - 5-15%   Manual work order system   _____

3. Time wasted by starting wrong priority work order
   - 5-10%   No work order system
   - 0-5%   Manual work order system   _____

4. Time wasted by equipment not being ready to work on [still inproduction]
   - 10-15% No work order system
   - 0-5%   Manual work order system   _____

5. Total of all percentages of wasted time (1 + 2 + 3 + 4)   _____

6. Annual Labor Costs   _____

DEVELOPING THE TPM IMPLEMENTATION PLAN

7. Total Wasted Labor Dollars (5 X 6) _____

8. Select the percentage from the table below that best describes
   your maintenance organization _____
       75-100% No work order or inventory system
       50-75%   Manual work order system
       30-50%   Manual work order and inventory system
       25-40%   Computerized inventory and manual work order system

9.   TOTAL SAVINGS (7 X 8) _____
   This amount represents the projected savings from labor productivity

Part #3 INVENTORY AND STORES SAVINGS

1. Estimated total maintenance inventory _____

2. Estimated inventory reduction _____
       15-30% No inventory system
       5-15%   Manual system

3. Estimated one-time inventory reduction (1 X 2) _____

4. Estimated additional savings (3 X 30%) _____

5. Total Savings (3 + 4) _____

Part #4 MAJOR OUTAGE AND OVERHAUL SAVINGS

1. Number of major outages and overhauls per year _____

2. Average length (in days) _____

3. Cost of equipment downtime in lost sales
       Use daily downtime rate times total days of outages
       (total must be in downtime cost per day) _____

4. Total estimated cost per year (1 X 2 X 3) _____

86   TOTAL PRODUCTIVE MAINTENANCE

5. Estimated savings percentage
   5-10%   No computerized work order system
   3-8%   Pert System
   2-5%   Pert System and inventory control system   _____

6. Total Cost Savings (4 x 5)   _____

TOTAL COST SAVINGS

1. Equipment Downtime Costs   _____

2. Labor Savings   _____

3. Inventory Savings   _____

4. Major Outage and Overhaul Savings   _____

5. Total Annual Savings   _____

Additional Areas of Savings to Consider

1. Warranty Costs   Warranty costs may be recovered from vendors for work that is done on equipment while it is still under warranty. These costs may or may not be recoverable if your plant personnel made the repairs.

2. Quality Costs   Quality costs are directly caused or could have been prevented by good maintenance practices. It is best to consult the quality control department to find what percentage of all quality problems are maintenance related. This amount times the total scrap and rework costs for the plant for the year can be a sizable dollar value. The projected improvements and the related savings can then be calculated.

3. Purchasing Costs   These costs are premium costs paid for short delivery notices on emergency orders. They may include expedited air freight, overnight delivery charges, or courier deliveries. Disciplined planning programs combined with preventive maintenance can dramatically reduce these costs.

DEVELOPING THE TPM IMPLEMENTATION PLAN 87

4. Overtime Reduction In many plants, maintenance overtime exceeds 20%. Plants with good maintenance controls will reduce this number to below 5%. This area can present opportunities for some significant maintenance labor savings. However, a reduction in downtime for the equipment may also reduce the production overtime required to make up the lost production. This savings may be substantially higher than the maintenance costs.

5. Energy Cost Reduction Properly maintained equipment requires less energy to operate. Some studies show this savings could be 5% of the total energy consumed by the plant.

Convincing the Corporation

This section of the program deals with winning and maintaining management support. It involves following these steps:

1. Developing the plan
2. Calculating the costs of the plan
3. Calculating the savings from the plan
4. Developing the cost/payback analysis
5. Presenting the program

If you have followed the outline of this book, you can accomplish the first four steps. Step 2 deals with the specific labor and supply (tools and materials) costs to put the various steps of the plan in place. More details on some of these costs will be found in subsequent chapters. Meanwhile, the rest of this chapter will concentrate on the fifth step in which the maintenance manager must be able to present the program in a bottom-line oriented manner.

It has been said about quality and is true for maintenance, that "Good maintenance is not hard to do; it is just hard to sell." Selling is the area where most programs to improve maintenance fall short. Effective selling is built on finding a need (increased profit, better quality, greater equipment effectiveness, etc.) on the part of the customers (management) and convincing them that your product (equipment maintenance management – TPM, and ultimately capacity) meets that need. In order to accomplish this, you need to understand all you can about your product and the customer's need. Then it is your responsibility to use this knowledge to make the customer see the solution.

The maintenance manager is generally the individual who must start the sales process. There may be others in the company as well, but the maintenance manger must lead the sales process if maintenance improvement is to achieve long-term success. Remember that TPM is not a program-of-the-month, but rather a long-term commitment.

Once the maintenance manager is committed to improvement, it is important to look at the long-term goals of the company and see where maintenance can complement and support these goals. The following steps can be used to help the maintenance manager accomplish this.

1. Educate your Management Documentation such as letters, articles, case studies, and benchmarks from both outside and inside sources are important tools for educating management. Communication should always be positive and related to costs-benefit analysis, including increased profitability and facility benefits.

2. Document Savings Any changes or improvements that are implemented should be monitored and cost savings should be documented. The benchmarks that were established in the initial analysis can be used to confirm the savings and improvement.

3. Think Results Always evaluate the results of any changes that are implemented. Measure these results against the goals that have been established. Even if the goal is not totally achieved, note the progress. Some benefit is better than none.

4. Always Look for Ways to Make or Save Money This step will always get management's attention. Any suggestions, even with slight improvements or savings, will be perceived as a positive indicator by management. This attitude always makes a difference when management is asked to invest in new maintenance projects.

5. Write Articles and Speak to Groups Promotion provides positive reinforcement. It helps to stimulate others to action within the organization. Public speaking to groups outside the company can also

DEVELOPING THE TPM IMPLEMENTATION PLAN 89

attract new customers and clients. There is little more rewarding to management than to learn that a major new account was influenced by meeting or listening to a knowledgeable member of the company.

6. *Keep yourself and other current* Staying current in techniques and technology has several benefits. First, management respects professional and knowledgeable people who can help them (the managers) be perceived as team members and contributors. Also, managers who have kept their knowledge current find it easier to implement changes and can do so more quickly than if they need time to get themselves and the organization up to current techniques.

7. *Know Your Numbers* Always know your financial numbers. If you do not work from facts, you will never be successful. Working with the financial and budgeting managers can truly enhance understanding of numbers and allow maintenance managers greater flexibility in preparing proposals for maintenance improvement projects.

8. *Be a Change Master*

It is important to stay close to the cutting edge in techniques and technologies. If you want to make progress, you need to make changes. Resistance to change is a difficult obstacle. Adapting the maintenance managers' attitude to one that is receptive to change makes it easier to change the organization. Any world class organization must be able to make continuous and rapid improvement. Speed is an important attribute.

Although many of the tips and techniques described in this chapter are basic management concepts, they are among the most overlooked items in maintenance organizations today. Only by concentrating on these basics will management support ever be gained and maintained. If the education and support of management is not given priority, the program will slow down within a year and die shortly thereafter. It cannot be overstated: "Gaining management support is the single most important step to beginning a successful Total Productive Maintenance process."

CHAPTER 6

PREVENTIVE MAINTENANCE

This chapter discusses the techniques that are most commonly used by maintenance departments to ensure equipment reliability: preventive (PM) and predictive (PDM) maintenance programs. The preventive and predictive maintenance programs increase the availability of the equipment. Even more important, they reduce the amount of reactive work the maintenance department will have to perform. This reduction allows the department to become more proactive and less reactive. The maintenance department can then become more controlled because they will no longer be limited to serving as firefighters.

Beyond the relaxation of pressure on the maintenance department, the results of the first part of the Overall Equipment Effectiveness formula begins to show positive change. The decrease in both breakdowns and the resulting downtime increases the overall equipment effectiveness, producing the following benefits:

- Support for Just In Time (JIT) initiatives
- Support for Total Quality Management (TQM) programs
- Creation of an environment for successful Total Employee Involvement (TEI) programs
- Reduction in investment in Capital Equipment Assets (CAE)
- Reduction of maintenance inventory levels
- Increase in corporate profitability

PREVENTIVE MAINTENANCE 91

How effective are the preventive and predictive maintenance programs in industry? As shown in Chapter 1, the majority of the companies are not satisfied with their preventive/ predictive maintenance programs. There are many reasons why this is true. The two most common reasons are the lack of management support and the failure to show results.

Reasons PM/ PDM Programs Fail

The lack of management support is the reason why the majority of PM/PDM programs are not successful. This problem has two parts. The first occurs if PM/PDM is not presented to senior management in a manner that ensures their complete support. Senior management may not have clearly understood the resources that would be required or the time required for the programs to produce results. The maintenance manager should have the responsibility for ensuring this understanding. In order for senior managers to understand what is involved in PM/PDM programs, they must clearly understand the following:

Initial costs for equipment and personnel
Time required to service the equipment (release time)
Time required to show results

If senior managers are presented with an unrealistic program, the PM/PDM programs will never succeed. As previously stated, to be successful, PM/PDM efforts take resources and time to show results.

Second, the results of the PM/PDM program are often not presented properly to senior management. Once PM/PDM efforts have been started, senior management must be kept aware of the improvements in equipment reliability. Performance indicators such as uptime increase, Overall Equipment Effectiveness increase, and PM/PDM costs vs. PM/PDM savings (cost avoidance) need to be tracked to ensure senior management support for the program. This support ensures that as equipment conditions improve, the true cause for the improvement (PM/PDM program) is highlighted.

The second most common cause for the failure of PM/PDM programs is the lack of actual results. This occurs when the PM/PDM program is designed. The efforts may have been set up for the wrong equipment or did not focus on eliminating the root cause of equipment problems. It would be analogous to a doctor treating symptoms, but never the disease. The problem may slow down or go into remission, but it will eventually resurface. It is then that comments like "I thought the PM/PDM program was supposed to stop this" start to be heard. It is a short step to the loss of management support for the PM/PDM program. Without senior management support, the PM/PDM efforts disappear quickly. When the program is in its initial development, it must focus on the true goal of ensuring equipment effectiveness. The next section outlines the steps that should be taken during the initial set up of a PM/PDM program.

Getting Started

The first step is to analyze how to start the program. There are generally two choices of how to start a PM program. They are by department (location) or by equipment. The department/location start means choosing the department or area of the plant that has the most problems with equipment downtime. This may be the bottleneck equipment or process of the plant. The reasons for the breakdowns should then be studied. Once the true causes of the breakdowns are found, the necessary corrective measures should be implemented.

The second method of starting a PM program is by equipment. This method involves taking a list of the top ten equipment items that cause the most downtime in the plant (regardless of location) and focusing the program on these items. The type of process, plant, or facility would determine which method should be used for a particular site.

In either case, the results shown on the equipment are the primary concern to plant or facility management. If properly documented, the results provide the necessary catalyst for management to become supportive of the program. Continued results will insure continued support, if the results are properly documented.

Types of PM/PDM Techniques

The most common type of PM/PDM techniques are listed in Figure 6-1. The discussion of each technique will help in selecting the proper technique to solve equipment problems.

PREVENTIVE MAINTENANCE 93

> **TRAINING ALTERNATIVES**
> 1. Routine—Lubes, Cleaning, Inspections, etc.
> 2. Proactive Replacements & Scheduled Refurbishing
> 3. Predictive Maintenance
> 4. Condition-Based Maintenance
> 5. Reliability Engineering

Figure 6-1 The most common types of preventive maintenance.

1. Routines, Lubrication, Cleaning, and Inspections

This type of PM is the front line of defense against equipment problems. It is also one of the most critical in development of a TPM program. If the operators are to perform maintenance tasks, they are the basic tasks mentioned here. These tasks require little materials and few tools.

Routines may be small inspections and adjustments that are performed each day prior to equipment start up or right at shut down. With a little training and coaching, these tasks can usually be performed by operations. The routines also include tightening. This task prevents looseness of any parts, which in turn prevents vibration and accelerated wear. The proper tools for tightening should be supplied to the operator and kept at the equipment.

Lubrication is another task that can be performed by the operator. The best method is to color code the lubrication point and the lubricant so that there can be no mistake about what lubricant goes to what point. It is also necessary to specify how much lubricant goes in each point of application. In many cases, too much lubricant is as bad as or worse than too little. The operator must be trained to know what and how much goes where.

Cleaning is one of the most beneficial tasks for a piece of equipment. While wiping off the equipment, the operator will spot many small problems that are in their early stages. The operator can then write a work request for maintenance to correct the problem before it becomes a major one. Cleaning also keeps contamination from starting problems.

Many equipment problems can be directly tied to contamination. Contaminated lubricant, dirt, or grit on machined surfaces will cause accelerated wear and failures. By removing the contamination and repairing the source of the contamination, problems can be corrected in their infancy. The Japanese stress cleaning the equipment on a continual basis. They also stress that the initial cleaning, when starting a program, reveals many hidden conditions that will eventually result in equipment failures. The importance of initial and continuous cleaning cannot be overstated.

Closely related to cleaning is the inspecting of the equipment. Inspections can occur during any one of the previously mentioned activities. The operator can spot the problems and request maintenance to make the repair before any capacity reduction or loss occurs.

While this process looks good on paper, there are several problems related to operations performing this part of the maintenance program. The first is training. All operators will have to be trained to perform these tasks in accordance with the standard practices and procedures. This training includes all related safety instructions and standards. Without the appropriate training, they will not perform the task properly, which will produce less-than-required results from the program and, potentially, the failure of the program. The hidden factor here is that very few companies adequately train their maintenance personnel. Yet the maintenance personnel are supposed to be the ones responsible for training the operators. If the maintenance personnel are not properly trained, then the operators' skill and knowledge level will be insufficient to produce the results necessary to keep the program effective.

A second problem is the current status of the preventive maintenance program in most companies. If you look at some current PM checksheets, what do you find? Are they complete? Are they accurate? In most plants, they are not. For example, if you have a PM checksheet for a chain (or belt) inspection, what does it say? Check the chain drive? That is what almost all plants have. However, it probably does not say what to check.

A good PM checksheet should list the items to check and give the specific parameters for the inspection. It should list tasks such as checking the deflection in the slack side of the chain and specify no more than 10% for the distance. It should list checking the

PREVENTIVE MAINTENANCE 95

sprocket alignment, the sprockets for hooked teeth, and the chain for unusual wear on the inside or outside of the link plates. The list can and should go on in considerable detail.

Most people like to stop about now and say "My people are journeymen and I expect them to know this" or "My people are journeymen, they wouldn't want someone telling them what to do, they know how to do it." Both of these arguments are worthless. First, the journeymen may be performing PM now, but they will not in the future. Operators, even with proper training, will need completed checklists. Second, the Japanese are big proponents of checklists. They have the philosophy that people can and do forget things. Checklists prevent this.

If checklists are properly detailed, they become a training tool to help individuals measure tolerance conditions. What is too hot? What is too long? What is too full? Knowing the tolerance can help both operations and maintenance personnel become more effective troubleshooters and, more important, trouble preventers.

The previous tasks make up the majority of the TPM tasks that operations personnel will perform. Remember, however, that the operations group never replaces the maintenance group. All they do is relieve approximately 10-40% of the current PM workload from maintenance. This amount is highlighted in Figure 6-2. Most of that

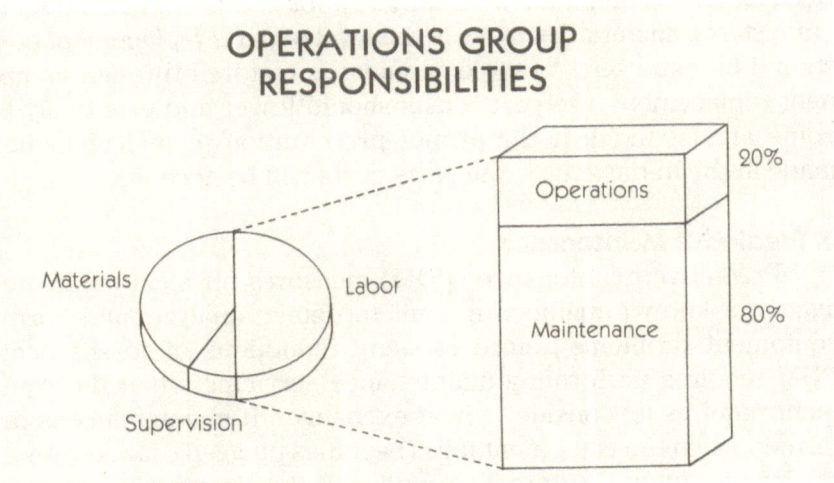

Figure 6-2 Examples of objectives for work order programs.

work is in the activities just listed. This transfer of work to operations does not result in a reduction of the maintenance workforce. Instead, it frees up the maintenance personnel to work on tasks and projects that they cannot presently handle.

2. Proactive Replacements and Scheduled Refurbishing

A proactive replacement is the term used to cover shutdowns, outages, and rebuilds. These activities involve taking a piece of equipment or unit off line and overhauling it, replacing all worn or suspect components, and then putting it back on line. The equipment will then operate for a specified time period, with little or no maintenance-related downtime. This technique is seldom used without support from historical records or predictive techniques. The proactive replacement targets equipment components that are approaching the end of their life cycle. This measure can only be determined by using the techniques just mentioned. If these techniques are not used, then the cost effectiveness of the program will be seriously affected. For example, if the parts costs are too high due to replacing components that are not worn, then the program will lose support.

One note must be made: Initial parts costs will increase when the program is initiated. The first few overhauls on the equipment will attempt to restore the equipment to its original condition. Because most equipment is usually in poor condition when the program starts, an initial increase in the level of parts replacement costs should be expected. As the equipment is restored through component replacement, the parts costs should lower and eventually become almost fixed. If the proper presentation to management is made in the initial stages, the parts costs will be accepted.

3. Predictive Maintenance

Predictive maintenance (PDM) measures physical parameters against a known engineering limit to detect, analyze, and correct equipment problems before capacity reductions or losses occur. PDM requires performing maintenance servicing when the equipment requires it. Consider a heat exchanger. If maintenance is performed when the equipment fails (becomes plugged), it takes longer to clean it out and restore it to service. If the cleaning is performed on a regular schedule—a fixed PM frequency—the cleaning takes

PREVENTIVE MAINTENANCE   97

less time and the equipment is out of service for a shorter period of time. If the exchanger is stopped up, the cleaning is much more difficult. However the problem remains "Are we performing maintenance too soon or too late?"

This question is answered by applying predictive techniques. In this example, the predictive measurement could be the pressure drop through the exchanger. If the pressure drop exceeds a certain limit, then it is time to clean the exchanger. The exact limit on the pressure drop can be established by the vendor, the engineering department, or even historical records. Once this limit has been established, then the cleaning occurs as needed. The frequency may be longer or shorter than the present PM frequency. However, the cleaning only occurs as required by the pressure drop.

Putting some estimates with the timing on the cleaning makes the benefits of predictive maintenance apparent. Suppose the cleaning when the exchanger is clogged takes 16 hours, the process costs $5,000 per hour, and adding in the maintenance labor and parts, the cost for the cleaning is $80,000 per occurrence. If the system fails six times per year, the total cost would be (6 X $80,000) or $480,000.

When a fixed frequency PM program is used, a certain interval is specified. If the failures occur once every two months, then the cleaning may be set for every month. The cleaning takes only 2 hours and the labor, material, and lost production cost is still $5,000 per hour. Now the cost is $10,000 per cleaning and the annual cost is (12 X $10,000) or $120,000 per year, an annual savings of $360,000.

If a predictive technique such as the pressure drop across the exchanger is used, the cost is still $10,000 per cleaning, but the exchanger may only need to be cleaned 7 times per year. This translates into a cost of $70,000 per year. The predictive method has a tremendous cost advantage over the failure method. It also has a considerable cost advantage over the preventive method. In this case, the only equipment required is a pressure gauge on each side of the exchanger. This expense is minimal compared to the savings.

The key to the predictive method is finding a physical parameter that will trend the failure of the equipment. The trend then can be used to predict the failure. Once the parameter has been found, it should have an upper and lower limit set, just like the upper and lower control limits in a quality control program. With the limit set, the current condition of the parameter measurement can be monitored by taking visible readings on a route. When the condition exceeds the upper control limit, then the required service is scheduled and performed. All of these actions take place before a failure occurs.

What are some of the major parameters that must be used to monitor equipment condition? The most common are listed below:

    A. Vibration

    B. Shock - Pulse

    C. Temperature

    D. Oil Analysis

    E. Resistance

While it is beyond the scope of this book to discuss these techniques in any detail, a short description of each follows.

A. Vibration Analysis

This technique monitors the frequency of vibration in rotating equipment. By choosing the proper frequency, all components in rotating equipment can be monitored and evaluated. All degradation of the component's condition can be trended and evaluated. Unusual wear can be detected and repaired before a failure occurs.

B. Shock - Pulse

This method detects the mechanical shock caused as equipment rotates. As the equipment deteriorates the level of shock impulse increases. The level of shock can be monitored. As it starts to reach a critical level, the equipment can be taken down and the defective component replaced before the failure occurs.

C. Temperature

This technique utilizes either the thermographic cameras or infrared scanners. These units monitor heat. As the temperature of a unit changes, the image generated by the scanner changes. This change alerts maintenance technicians to hot spots which, particularly in electrical equipment, indicate a potential problem. The technicians can then make adjustments or corrections before a failure occurs.

PREVENTIVE MAINTENANCE

D. Oil Analysis

This technique utilizes analysis of both oil and wear particles to determine wear. It can be used on any system that contains oil, such as mechanical drives or fluid power units. The analysis can check the condition of the oil itself, insuring that it has all its original properties. The analysis can also check the wear particles found in the fluid. The wear particles indicate that bearings, gears, and pump rotors are experiencing abnormal wear. This indication allows corrective action to be taken before a failure occurs.

E. Resistance

These checks allow the condition of key electrical components to be tested to insure that they are not shorted out or otherwise damaged. Developing problems in motor coils and other resistive components can then be isolated, allowing for proactive replacement of components before a complete failure occurs.

4. Condition Based Maintenance

Condition based maintenance is very similar to predictive maintenance, with the exception that workers are no longer sent out with handheld devices to take measurements in the field. The sensors are mounted permanently on the equipment and the signal is hard wired into a control room. Here all readings can be remotely checked, monitored, or perhaps integrated into a control system or a computerized maintenance management system.

Condition based maintenance will still utilize some or all of the techniques that are mentioned in the predictive maintenance section. With the enhanced usage of the programmable logic controllers (PLCs) and distributed control systems (DCS), real-time measurements from the systems in the field can be sent to the control room. Here the maintenance technician can monitor readings and actually perform some troubleshooting. For example, in a press and die operation, it is possible to monitor all hydraulic operations. If the system is properly designed, the maintenance technicians can monitor and trend hydraulic pressures during the clamping and pressing operation. If they spot a fall off in pressure during the clamping, and there is no leak, they will recognize an internal leak in the system. By further check of each area in the system, they can isolate a cylinder or valve leak.

When examining all areas of a preventive maintenance program, we quickly see that a good program utilizes all of the tools and techniques. Concentrating too extensively on one technique will produce a program that is not cost effective. The company that uses more diversity in the techniques will be able to design a cost effective program.

5. Reliability Engineering

Reliability engineering is a subject that can take several volumes of material to cover properly; this section will provide only a brief sketch. Reliability engineering is a discipline that is used to resolve problems that cannot be addressed by any of the above-mentioned techniques. It utilizes techniques such as redesign, retrofits, and even mathematical models to insure equipment availability. When problems continue to occur with equipment, the engineering group can become involved. Utilizing mean time between failure (MTBF) and mean time to repair (MTTR) calculations at a component level, the problems can be found in most cases and possible solutions uncovered. If a component is prone to failure, then it can be isolated and studied. If either materials or the design needs to be changed, the engineering group can make the necessary changes and then continue their evaluation to insure that the changes were effective.

STEPS FOR STARTNG A PREVENTIVE MAINTENANCE PROGRAM

♦ Determine Critical Units

♦ Classify Units Into Types of Components

♦ Determine P.M. Procedures for each Type of Compontent

♦ Develop a Detailed Job Plan for each of the Procedures.

♦ Determine a Schedule for each of the P.M. Tasks

Figure 6-3 The first steps to follow when implementing a preventive maintenance program.

Steps to Starting a Preventive Maintenance Program

Once the techniques have been analyzed and understood, it is necessary to begin putting the program together. How is this accomplished? The steps outlined in Figure 6-3 should be followed.

1. Determine Critical Units

This is the step mentioned earlier under gaining management support for the program. It involves choosing the equipment that will be included in the preventive maintenance program. Although the goal is eventually to provide preventive maintenance service to all equipment, the critical or problem equipment should be the starting point. Determining the critical units can be accomplished several ways, including:

The highest amount of downtime
The highest lost production costs
The biggest quality problems

The selection should be based on the criticality of the equipment, as specified by the operations or facilities that the maintenance organization services. This method insures that the correct equipment will be included in the program.

2. Classify Equipment into Types of Components

This step is designed to break the equipment down into its components. It is sometimes difficult to take a large piece of equipment and design a preventive maintenance program for the equipment in one step. Breaking down the equipment into components such as hydraulic systems, electrical systems, and mechanical drives makes it easier to design a program to maintain it. The more defined the components are in this process, the easier it is to design the programs.

3. Develop PM Procedures for Each Type of Components

This step will develop specific types of services that are required for each component, including all the techniques that apply to the components that were described in the previous section. These general descriptions of the services could serve as titles to the checklists and procedures that will be developed in the next step.

The information for the general lists of services that must be performed on each equipment item can be found in three basic areas:

1. Manufacturer
2. Equipment History
3. Operators, Craft Technicians, Supervisors

All three of these methods should be used and the results compared to insure full coverage for the equipment components. Each area has its strengths and weaknesses. The manufacturers will tend to overmaintain the components. The equipment history relies on the past to predict the future. Any changes in equipment design, operating levels, or maintenance tools and techniques can alter the component's maintenance needs. Operators, craft technicians, and supervisors all tend to rely on their memories and can forget or overlook items. If all three are combined, the correct maintenance services can be listed.

4. Develop a Job Plan for Each of the Procedures Listed

This step takes each of the individual lists of maintenance requirements and develops them into specific step-by-step job plans. This step is a critical and generally overlooked one in PM program development. The details of each PM inspection should include:

1. Craft required
2. Skill required for the craft (journeyman, apprentice, technician)
3. Tools required
4. Equipment requirements
5. Parts or spares required
6. Detailed job instructions
7. Estimated time to complete

Most of this list requires no explanation. The fourth point informs the craft technician of whether the equipment must be completely down before the inspection or service can be performed. If it is to be down during the PM, the details should also specify how long it needs to be down.

PREVENTIVE MAINTENANCE

The sixth point is one of the most neglected steps in U.S. companies. Most checklists or inspections are vague or incomplete. As previously discussed, without good details, items are overlooked or forgotten. If items are overlooked, failures or problems will result, causing a lack of program effectiveness. A PM program that is not effective will lose management support and eventually fail completely, spelling disaster for the maintenance organization and, ultimately, the company.

You often get only one chance to implement a PM program per company (or per career). Paying attention to these details will help insure that you do not fail.

5. Determine the Schedule for Each of the Tasks

Scheduling is a two-step process. It includes not only how long each task should take, but how often it should be performed. In determining how long it takes to perform a task, the estimate should consider:

- Time required getting tools and materials ready for the job
- Travel time to get to the job
- Any safety, environmental or hazardous materials restrictions
- How long it actually takes to perform the task
- How long it takes to clean up the area and put all tools and materials away

If good estimating techniques are used, then the scheduling and completion of the tasks is much more accurate. However, one factor can skew the schedules for PM programs—the time someone spends on the job performing work that is not on the PM inspection or service. As craft technicians service or inspect an equipment unit, they will occasionally find a problem that is beginning to develop. The question is how long should they spend correcting the problem before they ask for help, or write a work order to correct the problem. This is generally a policy decision that should be made when starting a PM program.

There are two factors to consider when making the decision. The first is the time it would take to come back to the dispatching point and write a work order to have the work done. Depending of the geography of the plant, the travel time could be considerable and should be taken into consideration. The other extreme is the damage a prolonged task would cause to the PM schedule. If a task is estimated for four hours, but takes eight hours to perform because of the other problems encountered, the schedule will suffer. Because the time is charged to a PM work order, performing what should be routine repairs and charging them to the PM charge number will inflate the PM costs, hiding the true repair costs.

In most cases, companies begin with a time limit of an hour. Any additional work could be performed up to the limit of an hour. If it requires more than an hour, then the craft technician should come back and write a work order to perform the work. This guidelines allows for planning and scheduling of the work, which should make it more effective.

This point leads into the second, the scheduling of the PM program. Scheduling PM tasks depends on what type of PM is specified. For example, OSHA inspections, environmental equipment, and hazardous equipment all require certain servicing on specified and regulated intervals. If these PMs are missed for any reason, and the regulatory government agency checks, the company would be liable for fines. These tasks are classified as mandatory PMs; they must be performed or something damaging to the company, equipment, or personnel will happen. They are usually fixed frequency PMs and cannot be altered.

A second type is the pyramiding PM, one that is due but is not completed in the allotted time window. As a fixed frequency PM, when it comes due a second time, another work order is issued. Thus, the work order pyramids. Non-pyramiding work orders are not issued a second time. They do not issue until the first work order is complete. With fixed frequency, PMs, the next due date slides based on the completion date, not the true next due date. Figure 6-4 highlights the danger associated with this approach.

With the pyramiding PM due on a fixed frequency, it is necessary to write a cancellation or missed notice on any uncompleted PMs. In this example, five completions would be noted, with two more being noted as missed. With the non-pyramiding PMs, there are also five completions during the same time period. However, when a failure occurs and the PM program is checked, the non-pyramiding PMs will

PYRAMIDING

	Jan	Feb	Mar	Apr	May	June	July
Due	1st	1st	1st	1st	1st	1st	1st
Completed	15th	Missed	1st	7th	Missed	20th	15th

Due = 7 times
Completed = 5 times
Number missed = 2

NONPYRAMIDING

	Jan	Feb	Mar	Apr	May	June	July
Due	1st	15th		5th	20th		1st
Completed	15th		5th	20th		1st	23rd

Due = 5 times
Completed = 5 times
Number missed = 0

Figure 6-4 Examples of objectives for work order programs.

show no missed tasks. This incomplete information leads the maintenance department to look somewhere else for the solution to the problem, when the real fault is with the PM program. Non-pyramiding PMs can hide potential equipment problems.

Another decision point in the program is whether to make the PMs just inspections or task oriented. If they are just inspections, the inspector must come back to the dispatch point and write the work orders for someone else to go out and perform the work. This can lead to a rift between the inspectors and the rest of the maintenance workforce. The task-oriented PMs instruct the inspector to make minor repairs and allow time for them to do this. When making this decision, the future must be kept in mind. If complex tasks are specified, it makes turning the PMs over to the operators more difficult. If the future direction is TPM, then the tasks must be set up and designed with that goal in mind. Otherwise the transfer of any of the maintenance tasks to operations will necessitate a complete re-write of the TPM program.

PM Program Indicators

How can you monitor the effectiveness of the PM program? How can you know when the program needs adjustment to insure effectiveness? Observations can be made in several areas:

1. Low overall equipment effectiveness (using the formula)
2. Longer MTTR (mean time to repair)
3. Maintenance-Related quality problems
4. Cost per repair increases
5. Rapid decrease in the value of capital assets

Low equipment effectiveness should be examined formula by formula. This could be particularly P.M. related when the availability is the lowering factor. If quality is the problem, then the quality part of the formula will highlight this point.

A longer MTTR (mean time to repair) indicates that a failure or breakdown takes a longer time to repair, meaning a more severe problem has been encountered. It takes longer to repair than a less serious problem that should have been found in its early stages by an effective PM program.

Comparing quality problems to the equipment effectiveness formula can help to spot those quality problems that are maintenance related. If the problems are related due to routine services or PMs, then the program's effectiveness is questionable. This indicator will allow corrective action to be taken.

The cost per repair is an indicator that shows repairs are more involved and taking more parts and labor than they should because of problems caused by a failure or an advanced stage of deterioration.

The rapid deterioration of assets simply means that the equipment and facilities are not lasting as long as their design intends. Lack of maintenance contributes to higher-than-normal capital expenditures for equipment replacements. If equipment is properly maintained, the effectiveness should be high enough to avoid purchasing replacements of backup equipment.

In developing any maintenance program, particularly TPM, it is essential to have a very effective PM program. This program eliminates the emergency or fire fighting maintenance that is costly in labor and materials. It also disrupts the relationship between operations and maintenance. It is imperative that the program be as effective as possible. If the PM program is ineffective, the rest of the program will suffer and eventually be discontinued. There have been many companies that have tried to replace their ineffective PM program with a TPM program. However, if they can't make a PM program effective, they will never make the TPM program successful.

CHAPTER 7

MAINTENANCE INVENTORY CONTROLS

If the previously-described steps have been followed to this point, you now have functioning preventive and predictive maintenance programs. Without these two programs, a company will never have the discipline to utilize inventory controls. Good maintenance practices require a disciplined approach to inventory and purchasing. Furthermore, an organization that does not have complete control of maintenance inventory cannot achieve the next step of improving maintenance effectiveness. Just as operations and facilities require a good maintenance organization to support their efforts, maintenance requires a good inventory and purchasing organization to support its efforts.

Common Delays

Poor inventory and purchasing practices are the single most common cause of poor maintenance productivity. What are some of the most common delays that inventory and purchasing can create in a maintenance department? They include:

1. Craft Technicians Waiting on Materials
Think of the time that your employees spend waiting to get the material to do their assigned jobs. This time can quickly add up to hours in just a single shift.

2. Travel Time to Get Materials

How much time do your employees spend going to the job and then finding they need to go back to the stores to get parts? Similarly, how much time do they spend getting the parts at the beginning of the shift? Do they have to wait in long lines at the stores window during the start of a shift?

3. Time to Transport Materials

Sometimes finding the material is only the first step. Transporting the materials to the job can take a lot longer and may involve finding a forklift or a truck to move the materials to the job site. This time can be even more costly when there a crew of workers is assigned to the job. Many will wait, while few make the move.

4. Time Required Identifying Materials

If stores materials lack either numbering schemes for identification or location schemes for finding the materials, considerable time can be spent simply looking for the materials. When parts are not numbered with a clear identifier, finding the correct parts can easily be confusing. One small difference in a part can render it unsuitable for its intended use. At that point, all the traveling and locating time begins again.

5. Time Required Finding Substitute Materials

Finding the right parts for a job is difficult enough, but when parts are out of stock, it becomes important to find substitutes. If other parts are not quickly identified as substitutes, substantial time can be lost in finding them as well.

6. Finding Parts in Alternative Store Rooms

As organizations grow; additional storeroom locations must be maintained in remote locations to reduce the amount of travel time. The problem develops of knowing what parts are carried or are in stock in each of these locations. If a part is out of stock in one location, how much time is needed to find out if it is in stock in another location? This information is important to prevent reordering the item when an adequate supply may already be on hand in a remote storeroom.

110   TOTAL PRODUCTIVE MAINTENANCE

> **TYPICAL STORES COMPLAINTS FROM MAINTENANCE**
>
> 1. I need this gasket but don't have the stock number.
> 2. I need a bearing like this. It is for the chiller pump on #1 water system.
> 3. These are the same gaskets but they have different stock numbers for each of the stores.
> 4. If we would have planned this job, all of the parts could have been ordered at the same time, which would have saved money, and they would have all been ready when we wanted to do the job.
> 5. Using so many vendors, it is hard to track their deliveries and prices.
> 6. I know we have items we could liquidate, but I don't have the time to look for them.
>
> Figure 7-1

7. Time to Prepare and Process a Purchase Order

Purchase orders can involve not just a considerable amount of time, but also a considerable cost. A crew of employees can waste a considerable amount of time waiting for a part to be processed through purchasing. This waste can be eliminated with proper controls.

8. Time Lost Waiting on Other Crafts

Inventory problems may be compounded in an organization that works with strict craft lines. If one craft has the materials to start its part of the job, but another craft involved does not, delays occur for the entire job and for all of the craft technicians involved. These delays can result in a tremendous amount of lost labor. If we compound these problems with others listing in Figure 7-1 the entire list of inventory problems becomes almost overwhelming. Therefore, inventory controls must be in place if maintenance effectiveness is to be achieved. In order to be effective with the inventory systems, the organization must understand how it should function and the information that should be maintained in the system. The following sections will address these points.

How a World Class Inventory Operates

In a world class environment, the planner is at the heart of maintenance's relation with the inventory department. The following work flow illustrates the planners' role.

1. When the work is identified, the planner determines what parts will be required.
2. Once the parts are determined, the planner issues the requisitions for the parts needed for the job so that they can be organized as a kit for the craftsmen.
3. The storeroom personnel gather the materials together in a staging area, which may be a bin or a palette. The parts are clearly marked with the work order number for the job.
4. When the work order is issued to the craftsmen, they go to the staging area and pick up the parts. Alternatively, the storeroom personnel may deliver the parts to the job site either the day or shift before the job is scheduled to begin.
5. The craftsmen use the parts. Any parts not in the kit may be requisitioned using the work order number. Any leftover parts are returned to the storeroom.
6. The storeroom personnel return leftover parts to stock and credit the correct work order with the returned part. The personnel should never be allowed to credit a work order for a part that was not issued to that order.
7. The items issued are compared to the items returned. This comparison is used as a measure of planning performance.
8. If items are not in stock, the demand for the item triggers a reorder. The work order causing the demand will be put on hold till the parts are received. The purchase requisition is sent to purchasing. In turn, purchasing consolidates the purchase requests into purchase orders, allowing them to get the best price on any item from the vendor.
9. A second method for triggering a reorder occurs when the number of parts is below the specified minimum for that particular item. This method is less reactive than the one described in the previous paragraph and should be set to avoid any stock outages.

10. Another class of item that may require reordering is the non-stock item. This item is one that management has determined is not necessary to maintain in stock. It may be one that only requires a short lead time, it may be of infrequent use with sufficient notice to order before use, or it may be subject to a contract delivery with a local vendor. The item is still tracked through purchasing, but may or may not be issued through stores, depending on the management philosophy regarding non-stock items.
11. Once an item is reordered, it must be tracked till received. The work can then be scheduled. If the item is overdue for delivery, this information should be noted and become part of the vendor's performance rating. Once the item is received, the planner should be notified so the work order on hold can be scheduled as soon as convenient.
12. Periodically the items in the store are counted to check for accuracy of the inventory. Any discrepancies are investigated and corrections are made. The quantities that are on hand should always be accurate to insure high levels of service from the storerooms.

Information That Needs to Be Maintained

This section looks at terms that apply to MRO (maintenance, repair, and overhaul) inventories. Some of the terms and definitions vary slightly from production-based inventory systems.

Inventory Information

On-hand quantity. This term is the actual quantity that employees would find if they walked to the storage bin or location and counted the number of the items that are physically there.

Quantity reserved. Also called committed to work orders, the quantity reserved is not the number of items that are in the physical location. Instead, it represents the number of items that are reserved for work orders already scheduled or ready to schedule. This number is critical because it lets the storeroom personnel know accurately how many items can be issued at the window.

Available to issue. The difference between the on-hand quantity and the quantity reserved is the available to issue. It is possible, due to incorrect stocking levels, to have a negative available to issue. However, most stock systems only take the quantity to zero. Therefore, accurate planning is important. Holding extra materials for jobs for which they may not be required may prevent completion of work that has a higher priority.

Quantity on order. The number of items that are presently on order, but not yet received, is the quantity on order. The due date also goes with this number, allowing the planner to note when the parts required are due to be received. If the parts are not received by their promise date, then the purchasing department should get a notice to follow up with the vendor to learn why the items have not been delivered.

Minimum and maximum on hand quantity. The minimum on hand quantity is the lowest number of items that you want in stock before an order is received. This quantity is different from reorder level, the point at when the order should be placed. In theory, the order should be placed before the minimum quantity is reached. If the minimum quantity is reached before an order is placed, a stockout is likely to occur. The reorder point should be set so that the lead time will allow the quantity to reach the minimum level just as the order is being received. What should trigger the reorder? When the quantity available to issue reaches the reorder point. This trigger is important because the reorder cannot be driven off of the quantity on hand. The reserves are deducted from the quantity on hand to get the true number available to issue. If the reorder is based on the on-hand quantity, then continual delays in scheduled maintenance work will result from the excessive stock outs.

The maximum quantity on hand should be the minimum plus one order quantity, which is the number of items ordered when a reorder is placed. This number plus the minimum will provide the maximum, which should never be exceeded. However, if poor planning disciplines or poor integration between a maintenance system and a stores system exists, this rule might be violated. When this occurs, then the company is keeping excess stores items and expenses such as holding costs, storage costs, taxes, and space requirements are all higher than necessary. Such costs are classified as inventory waste.

Last issue date. Another item to track in the inventory is the last issue date, which helps to monitor slow moving and obsolete items. Some large maintenance spares may never move. Still, it is good to review the inventory to insure that items with no movement within a specified time period are reviewed. This time period will vary from one company to the next, but usually falls between 6 and 12 months.

Physical count. If the inventory is kept in separate storerooms, such as in area or multiple storerooms, the stores personnel need the ability to query other stores locations to see if they have the part in stock before any orders are placed. If they do, then some companies perform material transfers, allowing any demands for materials made while there is stockout in one location to be met. Transfers reduce or eliminate downtime that would otherwise be incurred. A process called a physical count helps to insure that the quantities the stores are keeping are accurate. The stores personnel actually physically count the number of all stock items. The physical count is usually performed by location to keep the cost of travel low and enhance productivity. The actual count is then reconciled to the level the card system or computer indicates should be the count. Any adjustments that are made, either positive or negative, are charged or credited in accordance with the company's policy.

Cycle count. In a variation of the physical count process called a cycle count, stores items are classified, usually by amount of activity or by cost, and are assigned a code. For example, all "A" items may be counted once a year, all "B" items once every six months, and all "C" items once every three months. Once the frequency is established, then the items are divided so that an equal amount can be counted every week. This scheduling provides a smooth workload for the stores personnel and still keeps accurate track of the inventory.

Additional indicators. Some common performance information that should be kept for maintenance by stores include number of stockouts, number of issues, number of returns. Comparing the number of issues to the number of returns allows for a measure of planning effectiveness. Comparing the number of planned issues to the number of unplanned issues will also help to isolate planning problems. The number of stockouts could be used to identify items that have incorrect maximum-minimum levels and reorder quantities specified. Monitoring the number of adjustments can help spot pilferage or poor storeroom disciplines.

Purchasing Information

Purchasing information is a step beyond the storeroom data. The purchasing department must not only have all the part information, but also must keep vendor information such as the vendor contact's name, address, and phone number. Purchasing will also have to track the vendor's performance using measures such as number of late orders and number of orders shipped incomplete or incorrect. They may also track overages, underages, poor quality, and damage. All of these measures are then factored into a performance rating for the supplier or vendor.

The purchasing function has the responsibility to ensure that the prices for the items being purchased are the optimum available. Therefore, maintenance must avoid unplanned work in order to avoid dramatically affecting the pricing of an item. Emergency purchases block the ability to obtain volume discounts, price breaks on shipping, and planned or interval purchases.

If maintenance demands are planned and forecasted, then purchasing can consolidate all requests for items from one vendor into a multiple line item purchase order. This consolidation will also reduce the cost of purchasing items, a cost that is sometimes ignored by maintenance personnel. Consider the cost of processing a purchase order within your company. Is it $50? $100? $200? Depending on the company, single line item purchase orders can cost anywhere in this range to process. Multiple line item purchase orders can reduce this amount significantly. Thus, the logical action to reduce costs is to improve the forecasting and order policies for maintenance. Yet in many companies it is not uncommon to find purchase orders for items costing only $10, $20, or $30. Such orders are wasteful and must be eliminated.

Once these costs are controlled, what are some of the other areas to consider? They include the following:
1. Standardization of plant equipment
2. Standardization of supplies
3. Proper storeroom locations
4. Reduction and elimination of obsolete parts
5. Elimination of spoilage

Standardization of plant equipment. When plant equipment is standardized, maintenance gains inventory controls as well as additional benefits such as reduced training and reduced drawings. Inventory levels can be reduced because all of the equipment has basically the same parts. Instead of carrying 15 sets of spares for 15 presses, inventory may only need to carry 5 sets. Because maintenance repair and overhaul frequencies should be staggered to prevent simultaneous downtime, it is rare that more than a few sets of spares will ever be needed at any given time. These reduced levels can amount to tremendous savings for any size company.

Standardization of supplies. Just as standardization of plant equipment reduced parts inventory, the standardization of supplies prevents overstocking of similar supplies. For example, lubricants are one of the most overstocked supplies carried at any site. Consider how many different types and grades of lubricants are carried from different manufacturers, often more than twenty. Yet if companies go to their suppliers with their lubrication needs and ask for a sole source contract for all the plant's lubricants, the number might fall below 10 and, in some cases, even five or less.

Reducing the number of lubricants to that level eliminates many interchange charts. The number of lubricant-related failures from mixing incompatible lubricants will decline; many of these failures are presently going unsolved. Due to the chemical additives, the lubricants will coagulate or lose their viscosity when mixed. These problems can be virtually undetected unless oil analysis is performed. They can be compounded when poor training has been provided and the craft technicians think that all lubricants are the same. They can be further compounded when lubrication duties are turned over to operations personnel, who are less likely than the maintenance craft technicians to have the required training. This example with the lubricants is just a sample of the many problems that could be reduced or eliminated through part standardization.

Proper storeroom locations. The maintenance workforce must have quick access to proper storeroom locations in order to retain their productivity and prevent unplanned downtime. If standard supplies and critical spares are not stored in a location that is geographically close to the equipment for which they are needed, time will be lost due to travel as well as finding and transporting spares. In some cases, when storage close to the equipment is impossible, deliveries have to be made. However, deliveries only work well when good planning and scheduling disciplines are practiced. The problem further compounds in a true TPM environment, because all parts that the operators need are stored at the equipment. Without a delivery system, this approach would inflate the stores levels. Therefore, it is best to utilize area stores where necessary, then supplement them with a delivery system where it is not possible to have area stores. Some companies have eliminated up to ten percent of their total manpower expenditures by utilizing area stores and delivery systems.

Reduction and elimination of obsolete parts. A key element of delivering effective maintenance stores controls is to reduce, if not eliminate obsolete parts. Every spare part carried in stock has a penalty such as space, heat, lighting, and labor to move the part in and out of stock. These costs can be considerable. If 25% of the inventory is obsolete, then a considerable charge is being made to the company for parts no longer required. Obsolete parts are sometimes kept in the stores although the equipment has been retrofitted, sold, or scrapped. If accurate spares records have been kept, these spares should be removed or scrapped at the same time as the equipment, avoiding the collection of junk in the storeroom.

Elimination of spoilage. Buying too much of a spare or supply often leads to spoilage. In normal practice, this excess is an error on the part of purchasing. They may see the opportunity to get a volume discount on an item and order a larger volume. The item then sits on the shelf and spoils. It is important to use historical trends and patterns when purchasing maintenance spares. Volume usage of spares can vary dramatically from month to month. For example, the stocking level of an item may be 24, but maintenance has not used any in the last eleven months. Without investigation, however, it may not be discovered that maintenance only uses the item on a certain overhaul or retrofit. Further study may show that the retrofit occurs only once per year and, at that time, it takes all 24 items. Depending on when the inventory evaluation occurred, the decision may be to raise or lower the stocking level, both of which may be a mistake. If good planning is enforced, having a just in time maintenance inventory is best; spoilage will then be eliminated.

In order to help managers control the maintenance inventory, it is necessary to break the inventory into categories. The most accepted method is the ABC classification (See Figure 7-2). In the ABC classification, "A" items are those items that make up approximately

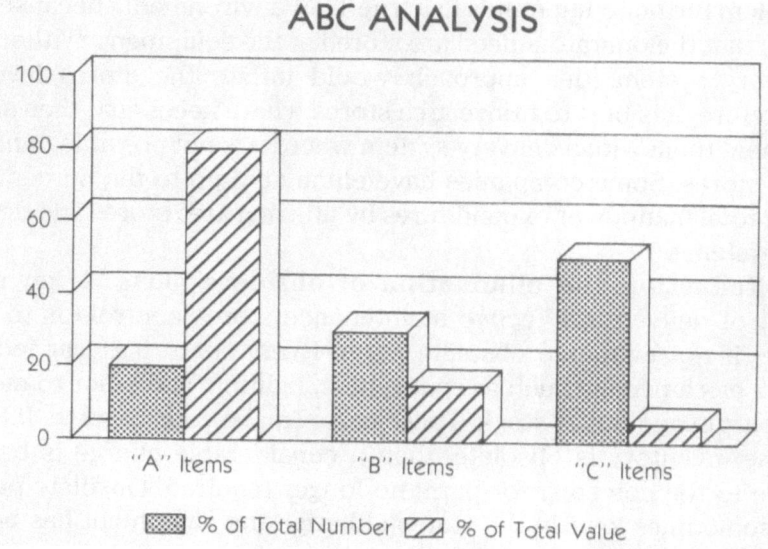

Figure 7-2 The "ABC" classification of inventory.

20% of the items in inventory, but are worth 80% of the total inventory valuation. "B" items make up another 30% of the inventory items, but are worth about 15% of the total inventory valuation. The "C" items make up the remaining 50% of the total items in inventory, but comprise only 5% of the total inventory value. Controlling the A and B items closely is a major priority because they make up the majority of the maintenance costs. C items can often be covered by blanket orders with vendors, allowing less labor and control expense to be invested by the company.

Maintenance Requirements

Basic Needs

What does maintenance require from an inventory system? Some of the basic needs include:
1. Real time parts information
2. Equipment "where used" listing
3. Projected delivery dates
4. Part usage by cost center

These needs are in addition to those already discussed.

MAINTENANCE INVENTORY CONTROLS   119

Real time parts information. Maintenance personnel need to know what parts are on hand at the exact time they are making the request. Some manual and computerized systems are updated only periodically. The update may occur once a day or, in some cases, once a week. Between times that the information is updated, the information is inaccurate. If a key equipment item is down and requires a part, someone could be sent to an area stores to get the part, while a crew is disassembling the equipment. When it is discovered that the part is not available, either the equipment stays down while the part is being ordered or it is pieced back together until the part is received. Either case wastes maintenance labor and creates unnecessary equipment downtime. To prevent this waste, maintenance needs to know what parts and quantities are in the storeroom at the exact time of the inquiry.

Equipment "where used" listing. It is important to maintain a list of all equipment in which a particular part or spare is used. This list facilitates the removal of all related stores items when a piece of equipment is sold, scrapped, or otherwise removed from the plant. It also allows for the planners to find equipment that uses the same parts or components and enables planners to check the service life of identical components installed on different equipment items. This information allows planners to check for failure patterns and, in turn, to determine if the failure is component problem or an equipment problem. This feature also allows for emergency borrowing of parts from equipment that is shut down or not in use.

Projected delivery dates. Planners need to know the projected delivery dates in order to project the possible start dates of work orders. In many cases, special order parts are required. Without these dates, scheduling work orders becomes guess work, which will damage the credibility of the maintenance organization. With good projected delivery dates, maintenance planning and scheduling becomes more accurate and reliable.

Parts usage by cost center. More for a financial audit, the parts usage by cost center can provide some useful trending information. If one department or cost area is using excessive parts, the maintenance practices and policies for that area can be investigated. If the parts are not being used, then the investigation can focus on how the labor for that area is being used. This analysis provides an interesting audit of maintenance policies.

Measuring Performance

As with all other parts of maintenance, the stores and purchasing functions need a method of measuring performance. Some of the most common indicators include:
1. Turnover
2. Stockouts
3. Service level
4. Inventory accuracy
5. Slow moving parts
6. Backorders
7. Purchasing service levels

Turnover. This indicator shows how many times the number of items in stock are used in a given time period. For greater accuracy, each category (ABC) should be evaluated separately. The number of turns will vary greatly depending on whether the items are A (high dollar, low volume), B (mid range value, medium volume), or C (low value, high volume). Companies that develop fixed rules, such as all items must turn three times per year, make a major mistake. This type of rule impacts the A and B items, the items that cause the highest amounts of equipment downtime when they are not available. The average number of turns for maintenance spares for the entire inventory should be one per year. C items will be higher; whereas A items may not turn during a particular given year.

Stockouts. The percentage of stockouts indicates the true level of inventory service. Stockouts occur when there is a demand for a part that is not currently available. It indicates a problem with the max-min levels, the reorder points, or another inventory set point. Stockouts can result in lost maintenance productivity and lost production. Both of these costs need to be considered when setting the order and stocking levels for parts. Stockouts should be less than 5%.

Service level. Just the inverse of stockouts, service level measures the percentage of time that the parts are carried in the stores. The service level should be greater than 95%.

Inventory accuracy. This measure compares what is actually on hand to what the records (computer or manual) indicate is on hand. The report can only be generated after a physical count of the items has been taken. The more accurate the inventory count is, the better control the company has over its inventory. Low accuracy shows that issues and returns being made to stock are not all being recorded. When this occurs, the stockouts and service levels will be affected. In order to be truly effective, inventory accuracy should be greater than 95%.

Slow moving parts. Stores items that are not turned within a specified period of time are considered to be slow moving parts. The time period, which depends on the company, ranges from 6 months to 1 year. This standard should be closely applied to all "C" items and some "B" items. By definition, most "A" items are slow moving. However, all slow moving parts should be evaluated at least annually to see if adjustments can be made that will allow for optimum stocking levels to be achieved.

Backorders. Vendor delivery performance and backorders are indicators of how the suppliers or vendors are doing in servicing the company's needs. The vendor should have a high on-time delivery rating, with very few orders either late or on back order. If the planners have scheduled work based on promised delivery dates, then the productivity of the labor force could be impacted. Vendors with low delivery performance and high backorder numbers should be replaced with more responsive vendors.

Purchasing service levels. Purchasing performance is measured by its service levels, which includes measures such as timely order cycles, updated delivery information, and rapid processing of transactions. Purchasing's responsiveness to maintenance needs is required to support the effort from stores. Without timely ordering, inventory levels will have to be maintained at an artificially high level to prevent stockouts and low service levels. All parts of the company impact one another in this cycle. If operations, which is the true customer for all of these departments, is not satisfied, then the profit picture becomes very bleak.

In conclusion: Just as maintenance exists to service operations, stores and purchasing exist to service maintenance. If this relationship is observed and understood, the company has taken a large step to insure its profitability and ultimately its competitiveness.

CHAPTER 8

IMPROVING MAINTENANCE EFFICIENCY AND EFFECTIVENESS

The techniques and controls described in this chapter build on the controls and disciplines from previous chapters. The ability to improve maintenance effectiveness is built on enforcing these basic disciplines. Without commitment to implementing the previous programs, the techniques described in this chapter will not be effectively utilized by any organization.

Examining Craft Technicians

Assuming the controls and disciplines are in place, you can begin improving maintenance effectiveness by examining your craft technicians at your particular site.

Do the craft technicians take pride in their work?
Is the quality level of their work high?

High levels of pride and craftsmanship are common in organizations where management supports the craft technicians. However, when examining your particular plant, do you find that management provides the craft technicians with:

The right tools for the job?
The proper materials for the job?
Materials properly located to eliminate lost travel time?

Equipment that is scheduled down so it can be repaired or serviced without a conflict with operations?

Clear job instructions detailing what is to be done and how it is to be performed?

Clear explanations as to the objectives and goals of the work assignment?

If these essentials are provided to the maintenance technicians, then they will take more pride in their assignments because they will feel like part of the team. Management will have provided the information necessary for the technicians to make logical decisions while performing the job. In this way, decision making is pushed to the lowest level. The job may be planned, but some decisions will still have to be made while the job is in progress. Good, clear information allows the technicians to make these decisions.

In most companies, inadequate planning techniques and the lack of coordinated information typically produce the following delays and wasted productivity for the craft technicians:

1. Waiting for or questioning instructions
2. Looking for supervisors to make a decision
3. Checking out the job because the instructions were incomplete or missing
4. Multiple trips to the storeroom because the parts were not planned or the wrong parts were planned.
5. Trying to find the right tools for the right job
6. Waiting for the approval to finish the job because it is going to take longer than originally thought.
7. Having too many or too few craft technicians, causing one group to look for more resources and the other to hide the excess resources they have.

In any of the above cases, the problem is lack of coordination and utilization of the available resources. These problems waste the available hands on time — the time the technicians actually have their hands on the job in progress. This is the time for which organizations actually pay. Labor productivity is one of the most important maintenance resources. How do other departments within a company control their assigned resources? For example, how do the financial, production, and engineering departments manage resources? Each one of these departments utilizes planning and scheduling techniques to maximize their effectiveness. If all these other areas use these techniques to optimize their resources, why don't maintenance departments use the same techniques to control their resources?

An important point often overlooked in maintenance management is that maintenance planning is the step that brings all the basic techniques previously discussed to this point together to optimize the utilization of all resources. Without maintenance planning, the organization will never be cost effective. Maintenance planning requires dedicated individuals to plan and schedule all maintenance activities. A good ratio of maintenance technicians to planners ranges from 15:1 to 20:1. If the numbers are higher than that, the effectiveness of the planning program is in jeopardy.

Planning Maintenance Activities

What is involved in planning maintenance activities? The planning is basically an effort to avoid all of the losses mentioned earlier. All work of a non-emergency nature, including PM and PDM inspections and services should have a detailed job plan. The job plan includes:

1. All required materials
2. The craft and skill specified
3. The required number of craftsmen
4. A listing of all required tools
5. A list of any non-standard equipment
6. A detailed description of all job steps
7. Descriptions of all safety requirements
8. Any OSHA, EPA, or other federal or state requirements

The maintenance planner's job responsibility is to insure that all of this information is provided for each non-emergency work order. This responsibility also indicates why the technician to planner ratio is so important. If planners are to provide all of this information, they will be limited to the number of work orders that they can process in a given time period. If the ratio is not kept reasonable, then the quality of the plan suffers and the effectiveness of the planning program becomes questionable. This problem has led to the ter-

IMPROVING MAINTENANCE EFFICIENCY AND EFFECTIVENESS

mination of the majority of failed planning programs in industry.

First-line supervisors in maintenance are then responsible for seeing that the jobs are executed according to the job plans provided by the planner. Supervisors must be on the floor with their crews at least six out of the eight hours available on the shift. They should not be tied to administrative or paperwork functions for more than two hours each shift. Otherwise, they become nothing more than high-paid clerks. Instead they must have proper control of their crews because they are responsible for their activities.

Scheduling brings the entire activity together. A maintenance schedule is most effective when performed on a weekly basis. Some companies try to schedule on daily, monthly, or some other time frame. However, they either lack the control or are too confining to be effective. The weekly schedule is the most effective because it allows flexibility, yet still has enough control to avoid wasting resources. For example, a daily schedule can be disrupted when made out 16 hours before the starting time of the schedule. Once the schedule is set, and the planners go home for the day, any breakdowns or emergencies that occur before they come back will disrupt the schedule. On a day-to-day basis, the schedule will be unreliable and inaccurate.

The weekly schedule is far more accurate because it allows for emergencies and other schedule interruptions. The amount of emergency work, small interruptions, PMs, and other interruptions can be tracked and averaged on a weekly basis.

Consider the following example:

10 workers X 40 hours	= 400 hours
2 contract employees X 40 hours	= 80 hours
5 O.T. shifts X 8 hours	= 40 hours
Total available to schedule	= 520 hours

Deductions

30% emergency work (.3 X 520)	= 156 hours
5% absenteeism	= 26 hours
PM work (20%)	= 104 hours
Total deductions	= 286 hours

Total to Schedule = Available - Deductions = 520 hours - 286 hours = 234 hours

If the planner schedules 234 hours of work from the backlog for the week, the crews have a high probability of completing the tasks. The same technique was described in the section on maintenance backlogs and staffing levels. If these techniques are used to set the amount of work to schedule, then scheduling accuracy of 95% or more can be achieved.

Scheduling Flows

How the maintenance scheduling process should take place depends on the organizational structure, but these steps should be followed:

1. The planners gather any uncompleted work that is outstanding at the end of the week.

2. The planners calculate the craft capacity for the next week.

3. The planners deduct outstanding (incomplete) work from the craft capacity.

4. The amount of craft capacity that remains is the total number of hours that can be taken from the craft backlog.

5. Based on priority, date needed, and equipment availability, the planners select the jobs from the backlog for scheduling. Any work that is to be put on the schedule for the next week must be ready to schedule. Therefore, all parts, tools, outside contractors, rental equipment, and other resources must be ready. Any work that is put on the schedule before it is ready will do nothing more than create lost productivity and wasted resources.

6. The planners continue to select work from the backlog, matching the labor resource requirements to the availability. Once the availability is matched, the schedule is complete. The planners may choose to list several additional jobs in a category of optional work. These are listed in the event that the number of emergencies are lower than expected or there is a change in production schedules, restricting the equipment that will be available for work during the week.

IMPROVING MAINTENANCE EFFICIENCY AND EFFECTIVENESS 127

7. The tentative schedule is presented to the maintenance manager for review and maintenance approval.

8. The maintenance manager should meet with the operations manager by Thursday of the week before the schedule is to start. For example, the operations manager may want to rearrange priorities, add specific jobs, or cancel selected others. This interchange between the managers should finalize the schedule.

9. The finalized schedule is presented to the planners who begin printing all work orders, parts pick lists (for the stores), contractor notifications, and equipment rental agencies.

10. The planners then put the information about each work order into a packet and deliver them along with the next week's schedule by noon on Friday to the supervisors who are responsible for the work.

11. The supervisors have time to look over the work schedule and resolve any questions before the end of the day on Friday. This allows them to prepare the sequence in which the work is to be executed during the following week.

12. A schedule does not tell the supervisor the sequence in which the work is to be done. This is the supervisor's responsibility. It is also the supervisor's responsibility to match craft technicians to the job. The supervisor is familiar with the skill level of each assigned employee. It is the supervisor's task to match each employee's skill level with the particular job.

13. As the week progresses, the supervisors will turn in all work orders that have been completed to the planners. The planner will complete the record keeping (This task may be a clerical one, depending on the resources available).

14. The planners monitors the progress of the schedule completion and by Thursday are ready to begin the schedule for the following week. The process begins to repeat.

Some variations and adjustments must be made to make this scenario work for all companies. Multiple planners will require that schedules are coordinated. Multiple crafts require the same additional coordination. Multiple operations managers require multiple meetings and maybe multiple schedules. Area maintenance will work somewhat differently than centralized maintenance. However, if the basic principles discussed here are applied, and the basic disciplines are enforced by all parties involved, good effective scheduling will occur.

By observing the requirements for good schedules, we can quickly see that no maintenance schedule will ever be effective if good planning is not enforced. Effective scheduling require that each work order have:

    Accurate craft requirements
    Accurate materials requirements
    Accurate contractor requirements
    Accurate equipment and tools requirements
    Accurate date needed, priority

Again, without effective planning programs, effective maintenance scheduling is just a dream.

A review highlights the following points:
1. All maintenance work should be scheduled.
2. All maintenance work should be planned by experienced technicians.
3. The schedule should use capacity planning techniques.
4. No work should be scheduled until it is ready.
5. The work should be processed as backlog, weekly, and then daily work.
6. The planner has responsibility for the backlog and weekly schedule.
7. The supervisor has responsibility for the daily schedule.

If the guidelines presented in this chapter are followed, then maintenance planning and scheduling can be successful. This will be true whether maintenance is scheduled by the maintenance department or as part of a team effort. Successful team-based maintenance organizations also utilize the guidelines presented in this chapter.

CHAPTER 9

MAINTENANCE INFORMATION SYSTEMS

This chapter examines the computerized maintenance management system (CMMS) and the ability to interface maintenance information to other organizational systems. The chapter begins with the benefits achieved by computerizing and concludes with the selection and implementation phases. In some TPM environments, the system is called an Equipment Information Management System (EMIS) because its focus is on managing equipment rather than maintenance. More recently, it has become known as an EAM (Enterprise Asset Management System) because the focus is more on managing assets' life cycles. With few changes, however, almost all of the acronyms are interchangeable with CMMS. For the purpose of this text, the CMMS acronym will be used.

GOALS FOR COMPUTERIZATION

- Provide Vehicle for Enforcing Maintenance Disciplines
- Provide Faster, More Accurate Record Keeping Capabilities
- Provide "Snapshot" Analysis of Maintenance Information
- Provide a Method of Integrating Maintenance with Other Information Systems

Figure 9-1

Why Computerize?

Figure 9-1 details the four main reasons for computerizing maintenance management.

Enforcing Maintenance Disciplines

Computerized Maintenance Management Systems provide a structure for enforcing maintenance disciplines. Every CMMS, whether purchased from a vendor or developed in house, has its own philosophy of how maintenance should operate. This philosophy governs how information is collected on a work order, stores and purchasing procedures, reports, and many other areas. The structure and philosophy of the CMMS will also determine how your company manages maintenance. Selecting the right CMMS, therefore, is very important.

At this point, the question "Which came first — chicken or egg?" comes up for the maintenance manager. In this case, the question is "Does a good manual system or paperwork system need to be in place before computerizing or does computerizing provide a good system?" Many consulting firms will answer yes to the first part, then spend months helping a company to develop the paperwork systems for managing maintenance. When the CMMS is implemented, many of these paperwork policies that the manual system put into place must be changed in order to use the computerized system. If you are going to use a CMMS, then implement it and enforce the disciplines that it requires, saving what can be considerable time and money. If the right system is selected initially, then developing a manual system first is a waste of time.

Improved Record Keeping

The CMMS should also provide a faster, more accurate record keeping methodology. Paperwork systems generally have only one point for data entry whereas a CMMS may have many points, allowing for faster record keeping. Because the systems are usually definitive on the data they require, the data going into the system must be checked for validity. This check insures the data is more accurate and complete than what is found in a paperwork system. The accuracy and timeliness of the data provides managers with better information on which to base their decisions.

MAINTENANCE INFORMATION SYSTEMS

Summary and Analysis Reports

The CMMS should also have the ability to provide snapshot or summary reports that allow a manager to see data in a concise, meaningful form. These reports eliminate the need to browse through pages of data trying to find a trend or even a particular fact. This feature is not available in some off-the-shelf CMMS packages; not all of them provide analysis or exception type reports. Some only provide lists, impairing the usefulness of the system.

Integration With Other Systems

A computerized system, if properly selected, also has the ability to integrate with other computerized systems in the company. These systems may include packages for existing stores or purchasing, payroll, general ledger, and production scheduling. The integration can eliminate passing paper or redundant data entry points. The savings in clerical support alone can be substantial. Data accuracy and reliability can prove to be beneficial in improving maintenance communications with other departments that may already be computerized.

The problem with CMMS is that in many organizations, maintenance managers are not involved in the selection process. In some cases, they do not know anything about the software until someone sets up the computer and software on their desk. In this case, the maintenance manager and the maintenance organization will not be prepared to use the CMMS. The mindset is that simply giving the managers a computer and software will make them more effective. This is not the case. Implementing and using maintenance software effectively takes time, practice and dedication.

What Makes a Good CMMS?

A good CMMS must support all maintenance activities. Therefore, it must have:
1. Work Orders
2. Preventive Maintenance
3. Inventory
4. Purchasing
5. Personnel
6. Management Reporting

132   TOTAL PRODUCTIVE MAINTENANCE

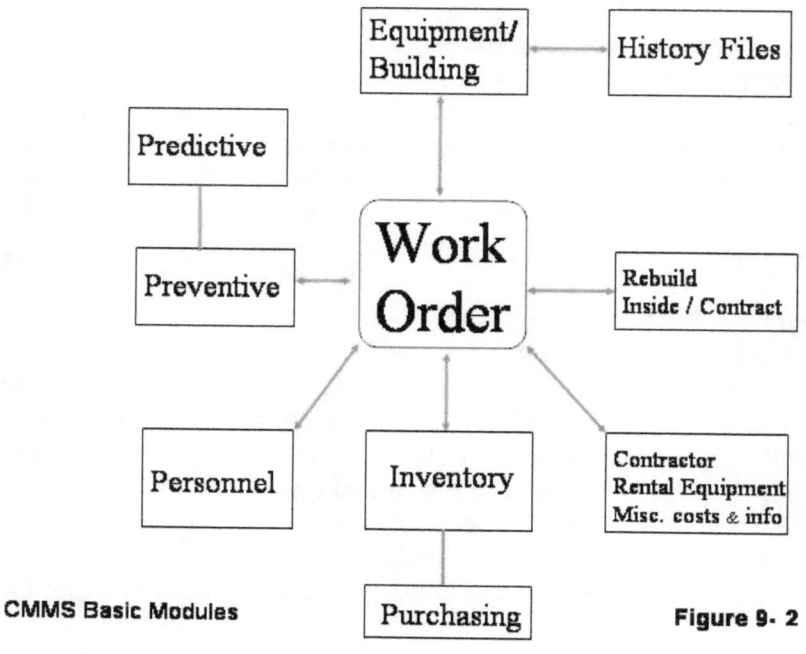

CMMS Basic Modules

Figure 9-2

The relationship of these modules is highlighted in Figure 9-2. Because the CMMS is an important repository for equipment data that will need to be used in a TPM organization, understanding the relationship of these modules is critical. Unless accurate data is input in all modules, the data stored in the equipment history will be corrupt and virtually useless for any decision making by TPM teams.

As has already discussed, all this information is required to make the maintenance organization function properly. Any CMMS must have these modules. However, just having these modules is not enough. How these modules work is the important consideration. For example, some systems can plan a work order in just a few keystrokes; other systems will require endless manipulation of screens and data to do a good job of planning a work order. Managers should examine and test a CMMS at this level before the final purchase is made. However, each organization should keep its needs in mind as the software is evaluated.

MAINTENANCE INFORMATION SYSTEMS   133

Vendors

If a packaged CMMS is considered, with what type of vendor can a company expect to do business? Vendors fall into five basic groups. They are:

1. Software Developers
2. Non-Maintenance Consultants
3. Maintenance Consultants
4. Maintenance Integrators
5. ERP – Financial System Suppliers

Software Developers

Software developers are programmers that have developed a software package for maintenance. They usually have little, if any, maintenance experience. Their software often reflects this inexperience; although it is usually very well written, with good coding practices, it generally lacks a definite maintenance philosophy and, in many cases, necessary maintenance disciplines. The vendor is usually very willing to take its base package and customize it to meet the company's needs. This process can be very expensive, however, and should be considered as a last resort.

Non-Maintenance Consultants

Non-maintenance consultants are firms that may already have production, accounting, inventory, purchasing, and other backgrounds, and that have decided to expand into the maintenance area. What are some of the signs of this expansion? Their package will reflect their area of expertise. If it is production control, the package will lack the flexibility that is required for maintenance planning and scheduling. If the background is financial, then the system concentrates on the financial aspects of maintenance. It will lack the necessary functionality to be effective in managing maintenance.

Maintenance Consultants

Maintenance consultants that have a package will generally enforce the necessary maintenance disciplines. However, the coding for the software is generally poor. The software will not take full advantage of the programming language or the hardware platform. Because these groups evolved from maintenance consultants, they tend to try making projects out of each sale. They may even try to put staff people on site for extended periods of time. Although this is not true for all companies, it is an area to watch.

Maintenance Integrators

Maintenance integrators are the full service vendors in the marketplace. They have the blend of maintenance and software expertise that allow for the premium products in the marketplace. They are often referred to as best of breed vendors (BOB). They can provide assistance during the implementation process in the area of systems and maintenance technologies. These are the best vendors to work with in the marketplace. Their major disadvantage is cost; they are relatively more expensive than the other vendors. However, sometimes you get what you pay for.

ERP - Financial System Suppliers

The ERP-Financial system suppliers have fully-integrated packages that have a maintenance module. These vendors offer the advantage of providing a corporate system that meets the needs (not necessarily the wants) of a company, with the advantage of having only one vendor to work with when problems arise. Issues around product upgrades, integration issues, version control, and system life cycle cost are minimized. Previously, these systems lacked some of the functionality required by the maintenance organization. More recently, these vendors have upgraded the functionality of their maintenance modules and now have virtually the same functionality as the best of breed vendors.

Selection Process

Selection Methods

What is the best method to select maintenance software? The process should begin with a company defining its needs. Begin by asking what you want the system to do for maintenance. Then expand the question into other areas of the organization that will be affected by the CMMS. Once these needs are listed, develop them into a requirements document. This document may be something as sim-

MAINTENANCE INFORMATION SYSTEMS 135

ple as a checklist. The document can also expand into a request for information, a request for proposal, or a request for quotation. Keep in mind, however, that the more extensive that your requirements are, the more the system is going to cost.

A major mistake that many companies make is to look for main-

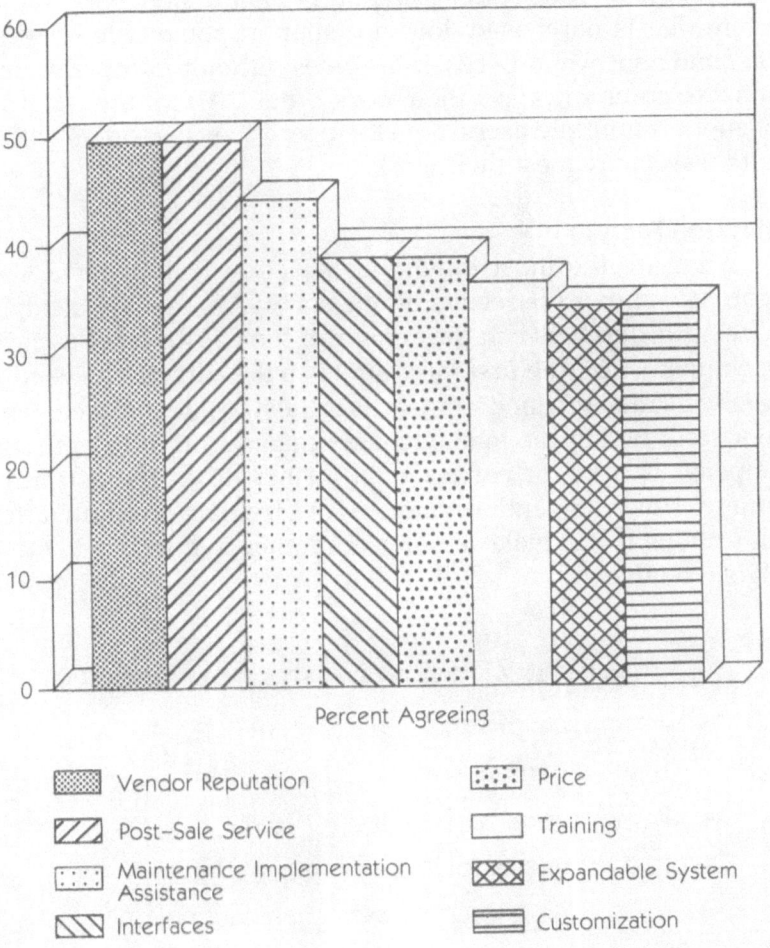

Figure 9-3 The most important considerations in selecting a computerized maintenance management system (CMMS).

tenance software without knowing what they want the software to do for them. When this occurs, the wrong software is purchased. The company will then expend unnecessary resources trying to make the software match the organization or the organization match the software. The endeavor becomes long and ultimately frustrating, never realizing its potential. Always know what your requirements are before looking for software.

Knowing your requirements becomes especially important if the end goal of your entire program is TPM implementation. If the system that is purchased does not support some type of work request and approval process, it becomes difficult for operations personnel to enter and track their work orders. Also if the CMMS does not support multiple users, then the operations personnel will never get to use it to request their work.

Selection Factors

What are the most important selection factors for a CMMS? Figure 9-3 shows the results of one survey. As can be quickly seen, the relationship between the vendor of the CMMS and the purchasing company take the first three spots on the survey. The vendor, especially its maintenance support, is critical to gaining acceptance of the system within the maintenance organization of the purchasing company. This factor carries even further down the list into user training. How the vendor conducts the training and how effective that training is will make a major difference in how quickly the system is effective.

SELECTION AND IMPLEMENTATION PITFALLS

- Wrong type of vendor doesn't meet needs
- No management support
- Wrong software/hardware
- No after-sales support
- Lack if willingness to make organizational changes
- Chooosing in-house development

Figure 9-4

Selection Challenges

What are the most common types of problems encountered during the selection and implementation of the CMMS? Figure 9-4 highlights the most common reasons. For example, it is possible to choose a vendor that does not fit your organization. When this occurs, either the vendor's software never meets the real needs of the organization or it requires too many support people to make it function. In either case, the CMMS does not function the way that was expected and the client company is dissatisfied. Then the client has a choice: to put up with the shortcomings of the software and vendor or to change and purchase another software program.

Management support is essential, but this hurdle should have been cleared earlier in the program. However, good communication will help upper management to understand what is occuring during the implementation. The lack of willingness to make organization changes can be a serious problem. In some cases, the organization needs to change to adapt to the maintenance system. No one has the perfect organization. Because a company is trying to improve by purchasing a CMMS, it should not be resistant to making changes that will help them be more effective utilizing the tool they have purchased.

In-House Development

The last point in Figure 9-4 is the thought of in-house development. Although many companies have opted for in-house development, it is a poor choice because they generally end up with one of two extremes. One is the system is never as fully functional as it could be, which means it does not do as much as it could for the maintenance organization. The other extreme is it is so complex that no one in the organization wants to use it. Companies considering in-house development should know that it takes twice as long to design and implement a system as it does to purchase one. Even worse, the development and internal support costs will be over ten times more than for a purchased package.

If a company cannot find anything on the CMMS market that meets its needs, it should negotiate with a vendor to purchase its source code, then customize it. At least the company will be starting with more than half of the system already in place.

The bottom line: in-house development is usually undertaken

STEPS TO SUCCESSFUL IMPLEMENTATIONS

1. UPDATE CURRENT RECORDS
  A. Equipment Nameplate
   Numbering schemes

  B. Preventive Maintenance
   Relate to equipment/components
   May have to write tasks

  C. Stores and Purchasing
   Numbering schemes
   Usage history
   On-hand quantities and locations

  D. Personnel Histories
   Current skill levels, etc.

  E. Work Order Files
   Accurate history

2. SYSTEM INSTALLATION
  A. Hardware
   PC protection
   PC/LAN system protection
   Backups
   User security
   Terminal locations
   Printers
   Mini/Mainframe

  B. Software
   PC/LAN
   Mini/Mainframe
   Understand the process

3. DATA ENTRY
  A. Electronic Transfer
   Cost from vendor

  B. Manual Information (Step 1)
   Forms
   Temporary clerical support

  C. Consistency
   Upper/lower case

4. SYSTEM INTRODUCTIONS
  A. Ongoing
   Pay envelopes
   Bulletin boards
   Crew meetings
   Supervisor contacts
   Company newsletter

  B. Characters
   (Nonthreatening)

5. PERSONAL TRAINING
  A. Right Staff People
   Don't train plant manager

  B. Right Type of Training
   Job/task oriented
   Hands-on, real world

  C. Right Conditions
   Your site
   Vendor's site
   Neutral ground

Figure 9-5 The Five Step Process

by a company or individuals within a company with an ego problem. They believe that they can do it better than any of the three hundred or more vendors with CMMS packages on the market. This ego problem has caused more problems and cost more money to companies than anyone would dare to admit.

System Implementation

What is the best method for implementing CMMS software? It is generally a five-step process.

Preparing and Updating the Data

The first step, shown in Figure 9-5, involves preparing all of the data for entry into the system. Equipment and inventory numbering schemes were discussed in earlier chapters, but now they must conform to the limits of the software. If you have a 20-digit equipment number, but the software only accepts a 10-digit number, you have a problem. This area is one of many that must be carefully considered during the selection process. In order to insure that the software matches the data, ask the vendor to give you the fields and their sizes, then develop forms to gather the data. If the forms match the format that appears on the screen for data entry, you can save time and money in the later stages of implementation.

Installing System Hardware and Software

As Figure 9-5 indicates, the second step, critically important to the success of the program, is installing the hardware and software. Many companies mistakenly believe that installing the hardware and software is as easy as plugging in the computer. In an industrial setting, power is unclean, subject to fluctuation and frequent outages. These problems will create data corruption, software failures, and general user headaches with the entire project. They can be avoided by using good protection devices on all computer equipment. Backup power supplies can be utilized in areas where frequent power failures occur. Industrial grade computers can be utilized in areas where environmental conditions are poor. Although somewhat more expensive, they help avoid loss of a system or the data, which are much more expensive.

Backing Up Data

Frequent backup of data is an important protection against data loss. In some cases, a power spike—a crash of the computer hard drive—wipes out all data in the computer. Without backups, all this information is lost. However, with backups, the software and data can be reloaded, then the system is ready to run again. How often should the systems be backed up? At least once per day, with intervals of at least twice per day during the initial data loading process. Proper backup of data remains one of the most neglected areas of any CMMS in operation, particularly in the PC or PC-LAN environments.

If the CMMS is going to operate on a minicomputer or mainframe computer, the company's information systems support personnel will probably be responsible for installation and backups. Nevertheless, the maintenance manager and those regularly involved in the use of the system should still know how the backups and system maintenance are scheduled and performed. This information will eliminate the black box mystique that these systems tend to develop.

Security

Most systems come with effective security levels that allow for multiple users on a system. These protections keeps data and some system functions restricted to those who need access to them. However, security is only as good as the users want it to be. If manager leave their passwords open where others can get access to them, then the security of the CMMS is compromised. By following the recommendations of the vendor, then adequate security should be assured.

Data Entry

The third step, as shown in Figure 9-5, is the data entry phase of the implementation.

Data Entry Clerks

If data has been gathered in Step 1 on forms that look like the system's computer screens, then you can hire data entry clerks temporarily to input the data. Using data entry clerks eliminates loading down the maintenance staff with this job. In fact most maintenance

MAINTENANCE INFORMATION SYSTEMS   141

staffs are already working hard just to keep up with their present workload; making them enter the data as well would take a long time. In turn, if data entry takes too long, then the enthusiasm for the CMMS project starts to wane. However, if date entry professionals are used, then the process is quick, without increasing the maintenance staff's workload.

Transferring Data

If the present data is stored in a computer, whether in another CMMS or a homemade database, the vendor of the new CMMS may be able to electronically transfer it from the old system to the new one. This transfer can save a tremendous amount of resources during the data entry phase. However, such a transfer will depend entirely on the vendor of the CMMS as well as how you have kept your data in the present system.

Data Reliability

One additional point about data entry is important; Do not use any data that is suspected of being inaccurate or unreliable. The theory of garbage in - garbage out is very true when it comes to computer data. Make sure that all data that is entered into the system is both accurate and reliable; otherwise, none of the output from the CMMS will be.

Introducing the System

The next step is to introduce the system. Although this is listed as the fourth step in Figure 9-5, it actually is ongoing throughout the selection and implementation process. For some organizations, starting a CMMS program can be quite a cultural shock. Good communication can help limit the amount of resistance to change. One vendor has even been successful at inventing cartoon characters to help the employees identify with the system under the theory that it is hard to feel threatened by a cartoon character.

Other methods listed in Figure 9-5 can be useful for ongoing communication. Each site will have to determine the method it will use to keep the lines of communication open.

Training

Training, the final step of implementation, is one of the most overlooked areas of system implementation. It is an area that companies too often omit when submitting project budgets. This omission is a major mistake because training determines how quickly the system will be implemented and how fully it will be utilized. Yet despite many warnings, companies still believe that they can do proceed without training. This belief is absurd. Consider some of the most popular courses offered at adult education classes, those that teach the use of database and spreadsheet packages that cost under $1,000. These courses are almost always full. The software comes with user manuals, but the students still want to learn more or learn from others. They want to become proficient with the software and the only way to do this is through training. If individuals will spend their own money learning how to effectively use relatively inexpensive personal software, companies should certainly invest in teaching their employees how to use software that costs thousands of times more than personal software. Their resistance does not make much sense, yet companies continue to fall into this trap.

Training must also be the right type. It should address real-world problems, using the company's data. Trainees should have to use the software, not just listen to lectures. Lessons should be in a combination classroom/workshop setting, free from interruptions and distractions of the employees' normal jobs. In order for the implementation to be successful, training programs should be available to all who are going to use the software.

Additional Areas of Concern

Two additional areas of concern are underestimating the time to implement and overselling management of the benefits of computerizing.

Implementation Time

The time needed to gather data and enter it into the CMMS can be considerable. Most consulting firms estimate one hour per record to gather the data and enter it into the CMMS. The condition of the data will have an effect on the time; nevertheless, this rule of thumb is fairly accurate. Therefore, if a company has 10,000 inventory items, and 2,000 pieces of equipment, then the time becomes rather considerable. If it dedicates only one person to enter the data, years

> **WORLD CLASS VERSUS PRESENT STATUS COST BENEFITS**
> **WHAT TYPE OF SAVINGS CAN BE REALIZED?**
>
> **1. Labor** 50% increase in productivity
>   15-30% savings in labor costs
>
> **2. Maintenance** 15-25% reduction in inventory
>
> **3. Tools** 20% reduction in repair and carrying costs
>
> **4. Supplies and Miscellaneous**
>   15% reduction in maintenance expenditures
>
> (All figures are industry averages)
>
> Figure 9-5

will be needed to get the job done. The correct approach is to estimate the number of records, calculate the time needed to gather and enter the data, and then dedicate the resources to get the job done. This approach keeps upper management satisfied, when dollars and time frames are consistent and understood from the beginning.

Overselling the Benefits

Another problem is overselling management on the benefits of computerizing maintenance. Industry averages can help to some extent (see Figure 9-6), but each organization is different. The size of the savings and the improvements must be calculated with a full understanding of the present organizational conditions and constraints. As was seen earlier in the book, computerization alone will not solve some problems. Managers must understand the present condition of the organization and its future course before putting together a cost justification for a CMMS.

Summary

Although CMMS is a great tool for maintenance, it is just that: a tool. It is not a magic cure for all the problems that ails a maintenance organization. If managers keep this chapter in focus during the selection and implementation process, then they should have a successful project.

In connecting the CMMS to TPM, it must be emphasized that the data necessary to improve overall equipment effectiveness will be stored in the equipment history module of the CMMS. Without complete and accurate data input into the CMMS, the TPM effort will never fully mature.

CHAPTER 10

CAPACITY ASSURANCE TECHNICIANS

It has been said that to improve maintenance the first step is to change its name; otherwise, it will never receive any respect. Each company has to make its own name and acronym for maintenance, but as a sample, I use Capacity Assurance Technician or CAT for short. Although the idea may seem insignificant, it has made a difference in companies that have tried it.

Beyond just a title change, maintenance workforces are in need of education. This has been evident even more recently in view of educational statistics that show 13% of all U.S. 17-year-olds are illiterate versus only 1% in Japan and Germany. Furthermore, the graduation rate in the United States is only 73%, whereas it is 94% or more in Japan and Germany. These statistics add credibility to the claim that the American workforce needs more training. How effective will these people be when they enter the workforce? What added burden of training will be put on American companies?

In the matter of industrial training, consider how much training companies are funding at the present time. When was the last time that your craft technicians have been exposed to meaningful training? Six months? One year? Two years? Even longer? Most training groups will admit that if you have not updated your craft technicians' skills in the last 18 months, their skills are outdated. Some companies believe they are spending enough on training, but when these claims are examined, only 20% of the training dollars are spent on technical training. The rest is spent on compliance or "soft-skills" training. But how are training programs organized? What subjects should be included? The following sections will answer these questions.

Training Programs

The various options for filling the need for a highly-trained workforce are listed in Figure 10-1.

Hiring Already-Trained Workers

The easiest option involves no formal training, but instead is to hire workers who are already trained. These workers are usually available from other companies in your geographical area. What are some of the factors to consider when following this option? First, although these workers may be skilled, their knowledge will be general, not specific to your equipment, process, and organization. Even if the need for formal training is eliminated, there still is the need for some on-the-job training to help them adapt to their new surroundings and job assignments. Another disadvantage to this option is its expense. A good journeyman in any craft is probably earning maximum wages for the geographic area. To hire workers of any quality away from their present position will require a better opportunity (usually financial). The issue then becomes, "Is it more cost effective to hire the expertise at a premium cost or to hire someone at an ap-

---

TRAINING ALTERNATIVES

- Hire Trained Personnel
- VocationalSchools
- Train In-House
- Train at Vocational Schools
- Vendor Training
- Colleges and Universities
- Continuing Education Programs

Figure 10-1

prentice level and train them to become a journeyman?" The correct answer will vary from area to area, but the issue should be evaluated closely before any final decision is made.

Trained Apprentices from Vocational Schools

The second option is to hire trained apprentices from a vocational training program. Many vocational programs graduate good apprentice-level individuals who can quickly be productive. An advantage of this method is that the workers come to the job with theoretical knowledge accompanied by some hands-on experience in a lab-type setting. Their limited hands-on experience is not a real handicap because they have not had the opportunity to learn bad habits elsewhere. They will be able to start at entry-level compensation, yet still possess the basic skills needed to contribute to the work effort. This option helps keep initial employment costs low.

What are the disadvantages to hiring people with vocational training? Their knowledge is not complete or may not provide competency in some areas. The level of proficiency that they bring to the job will vary greatly depending on the background of their instructors and their programs. If their instructors approach is to focus on theory, then all these workers bring to the job is textbook knowledge. If the instructors provide real-world, shop experience, then the workers will have a higher level of job-related skills. Even then, this approach is not a cure-all because it still requires the use of an on-the-job or other form of in-house training program. In turn, if the in-house training program is not used to complete the training of the vocational employees, then years may be needed for these employees to learn by osmosis what they need to be effective.

In-House Training

A third option is in-house training, which occurs when the company sets up and maintains its own training center for maintenance. When all of the options have been considered, experience shows that this method is the best for training craft technicians. The negative factor is that it is also the most expensive. Even though it may be the most expensive option, however, it is the most controlled; it is targeted to meet specific needs and produces the highest quality employees in the least amount of time.

When establishing an in-house training program, several issues

must be addressed:
> Who does the training?
> What facilities are required?
> What materials should be used?
> Are the employees compensated for their time?
> What is the true cost?

Who Does the Training?

Depending on who you ask in the organization, this question will receive many different responses. Although there are many answers, the best teachers are generally crafts technicians or supervisors who have good job experience and possesses the ability to teach. They should have the respect of the trainees for their job knowledge and should teach the material in a manner that shows practical application. They can then make specific application of the material, avoiding the usual question of "Why do I need this information to do my job?" Enthusiasm usually runs high when students know that the material will enrich their knowledge base and ultimately their own self worth.

What Facilities Are Required?

The facilities needed for training include classrooms and labs. Classrooms should be used for lectures. They should have overhead projectors, slide projectors, video machines and, even in some cases, computers for any computer-based training. Adequate lighting, heat, and air conditioning are also important considerations. The classrooms should be set up with tables for writing and using reference materials. The labs should have equipment that can be used to highlight the material covered in the lectures. Being able to teach the theory and then use the lab to apply it in real-world action provides meaningful lessons to the trainees as well as the reinforcement necessary to make their retention rate high.

What Materials Should Be Used?

Appropriate training materials include textbooks, video tape programs, handouts developed either in-house or custom developed by outside companies, and, for remote learning away from the training site, correspondence materials . The best results are usually obtained by using a blend of these materials. Vendors for these materials are usually listed in any of the magazines and newsletters that service the maintenance market. They may also be found at the trade shows that have a maintenance focus. Each of these types of training materials has its advantages and disadvantages, but evaluating these would lead into a discussion of training philosophy and methodologies, a discussion beyond the scope of this text.

Are the Employees Compensated for Their Time?

Whether or not employees should be compensated for attending training varies from company to company, often with the collective bargaining agreement being the deciding factor. If employees are paid for attending these programs, they have an additional incentive to want the training. Yet it may be easier to implement training programs where the attendance is unpaid. Smart companies will realize that they benefit from the training as well; therefore, they will work with the employees to find an equitable solution. When companies have an agreement to pay for time on the job, it may be harder to get employees to want either paid or unpaid training. The entire company philosophy on training, pay for skill, and career advancement potential must be considered when investigating training compensation.

What Is the True Cost?

Without exception, quality in-house training programs are more expensive than any other form of training. However, if they are properly conducted, they are the most effective form. However, they require resource commitment from upper management, with the realization that there will be no short-term pay back. Classrooms, labs, equipment, and training materials will be a considerable expense. Yet the results from in house programs are faster and of better quality than any of the other methods.

Vocational-Technical Schools

Specific training programs can be set up and conducted at local vocational-technical schools. These courses use the Vo-Tech school's facilities, but use the material you specify. The course can be set up in several different ways. One may be just for your company. Another may be focused on a particular topic you want, but the course can be open to any companies that want to attend. The difference between the options is the cost. A course specifically for your company means that you will pay the entire cost of the course, minus any state or federal funds that are available to subsidize the cost. A course open to the general public will allow the cost to be spread over all the participants.

A question mark of the vocational classes focuses on the instructor. In some cases the instructor may lack the necessary real-world experience to be effective with your technicians. Instructors who have only theoretical backgrounds, but have not actually applied the material, will find themselves challenged frequently by the technicians. This may prevent a sufficient transfer of knowledge that is necessary to make the training program effective. Again, training the technicians requires an instructor with real job experience.

Vendor Training

With vendor training, either technicians can be sent to a location where the vendor already has a school or the vendor can come to your plant and conduct the course on-site. Concerns about vendor training include the question of the actual presentations. Sometimes, the vendors send technical instructors who do an excellent job conducting the training program. However, vendors also at times send sales representatives who slant the presentation towards a description of their products, instead of addressing technical aspects of using the products. The wrong instructor can turn a good program into a total waste of time and money. Always investigate the background of the presenter before the training program is set up.

Vendor training can generally provide good equipment or product-specific training. In most cases, the vendor's technical personnel are competent and very qualified. They should be able to answer most questions and use common examples to reinforce their presentation. The training materials are of good quality, in part because they often double as sales materials. Although this type of training usually has a narrow focus, it should never be overlooked for specific equipment or problem-solving.

Higher Education

Colleges and universities as well as community colleges can be used for some training programs; however, much of this training will be theoretical. For example, colleges and universities are good for teaching advanced electronics, fluid power, and mechanics training. When advanced courses are needed, these are often a good option. Many companies use this method to train their instrument technicians or electronic repairmen. The training is effective and can provide the background necessary for maintaining the advanced technological equipment in the modern plant.

Continuing Education

Continuing education programs often specific seminars addressing a variety of topics. These programs are usually two or three days in length and cost hundreds of dollars for someone to attend. If you need this particular program, it may be the only place to get the information without developing your own course. If you need many of employees trained, consider having the seminar leader conduct the seminar in-house for the company. The costs, which include the seminar leaders' fees and expenses, are typically more economical than sending several employees to a public seminar. In many cases, the seminar leaders can tailor their course to a company's specific needs.

Course Outlines

What topics should be covered in a good maintenance training program? Figure 10-2 lists several basic topics. For example, mechanics training should not include bridge stress calculations, but should involve subjects such as:

      Bearing care and maintenance

      V-Belts care and maintenance

      Chain care and maintenance

      Gear care and maintenance

      Screw threads and drives

      Mechanical fasteners

The list can become quite considerable in length. What is important is that training should teach employees what they need to know to do their jobs correctly. Some companies conduct a job needs analysis, which includes studying a particular job and seeing exactly what tasks are involved. Each task is then studied to identify what

> **TRAINING TOPICS**
> - Mechanics
>   - "How To"
> - Fluid Power
>   - Hydraulics
>   - Pneumatics
> - Electricity
>   - AC/DC
>   - Low Voltage
> - Electronics
>   - Board versus Component Repair

Figure 10-2 Examples of topics that should be taught in a maintenance training program.

knowledge is required to perform it. This required knowledge is then written and developed into a training program.

Consider the fluid power course in Figure 10–2. This course could be taught from a design perspective, making it theoretical, but not necessarily practical for the maintenance technicians. However, consider these subjects:

>   Fluid reservoirs: Care and maintenance
>   Flow control valves: Care and maintenance
>   Directional control valves: Care and maintenance
>   Hydraulic pumps: Care and maintenance
>   Troubleshooting fluid power components

These topics can also be expanded. What is important is to make the material job related, teaching enough theory to help the technicians understand the hows and whys of repair and troubleshooting.

Training in electricity requires the most coverage of theory of any of the industrial maintenance-related topics. However, even it can be simplified by teaching component functionality and operations. Circuit operation and troubleshooting techniques must also be included, but these topics help technicians see the application of the theory.

In electronics, the structure of the training program depends on what you want the technicians to be able to do. If you want them to be able to troubleshoot and replace components on a board, they need extensive training. However, if you just want them to be able to change the defective board and send it out for repair, less training is required.

All decisions regarding training content relate to knowing your needs. Therefore, a needs assessment or needs analysis is a prerequisite to developing a cost-effective training program. Otherwise, you could spend either too much or too little on training. Either one is a waste of resources.

Conclusion

Training programs can help with TPM's ultimate goal of transferring some of the maintenance tasks over to operations. If the training materials are developed properly, the maintenance technicians can then use them to help the operators understand the hows and whys of their tasks. If the materials are not correctly developed, however, the effort will have to be duplicated to develop the operators' training materials. More on this topic will be presented in chapter 12.

With technology changing so rapidly, training the maintenance and operations personnel is of ongoing importance. Companies that invest in training will be competitive. Those that choose not to train will eventually be out of business.

CHAPTER 11

TOTAL ECONOMIC MAINTENANCE

If the plan has been followed up to this point, then the maintenance organization is beginning to develop a large amount of data. The challenge becomes how to use this data to optimize the effectiveness of the maintenance function in the company.

Communication

By this time, the traditional barriers between maintenance and operations should be disappearing. The real source of this traditional, long-standing problem is the lack of communication. Each side has difficulty justifying to the other its own side of the discussion. In fact, operations managers may still have some difficulty in understanding how much priority should be given to maintenance activities relative to their own goals and objectives. This chapter discusses techniques that can be used to translate all factors into a common language, one that all the company can understand.

The search for language that is understood throughout the company can be reduced to one phrase: Financial Justification. Everyone must understand how maintenance contributes to the financial standing of the company. One statement must be clear to all involved:

Cost-Benefit Decisions

Nearly all maintenance decisions are cost-benefit decisions. They must address the question "What is the company going to get

TOTAL ECONOMIC MAINTENANCE

in return for making this investment on maintenance?" The answer may be any of several, including reduced depreciation of assets, increased production, or higher quality. The problem lies in expanding the point of reference in order to see the entire picture.

Maintenance Costs

The cost of maintenance can be broken down into four major categories:
1. Labor Costs
2. Material Costs
3. Overhead Costs
4. Function loss or Function Reduction Costs

Labor Costs

Labor costs are the costs of having the maintenance technicians on staff. They include all related expenses, including overtime and benefits. The labor cost of each job can be calculated as the number of employees times their hourly rate plus any incentives (and related expenses). This cost makes up approximately half of all direct maintenance expenditures.

Material Costs

Material costs include all parts and equipment replacement expenses controlled by maintenance. The costs of all parts or equipment components used on a job should be tracked through the work order system. Material costs make up about 40% of all direct maintenance expenditures.

Overhead Costs

Overhead costs include the clerical and staff support for maintenance. Typical overhead costs range from about 15 to 25% of the usual maintenance labor costs.

Function Lost Costs

Function loss or function reduction costs are the hidden costs that are difficult to ascertain. These costs will range from 2:1 to as high as 15:1 per cost of maintenance activity. Therefore, if a maintenance task costs $10,000 in labor and materials, function loss or function reduction costs could run the true total costs of the activity to as high as $30,000 to $160,000. This increase makes a tremendous difference when evaluating the cost of a maintenance action.

One problem of calculating this type of cost revolves around the accuracy of the data. How do we judge what is worth spending and when we should spend it? A second problem is being objective when gathering this data. Without reliable methods of data collection, the data may be distorted, favoring either the operations or maintenance viewpoint. It is important to collect the data accurately in order to insure that the company gets what it pays for when it comes to maintaining its assets.

Maintenance Benefits

The benefits or savings from maintenance revolve around cost avoidance, when spending money on maintenance avoids incurring a greater equipment repair cost. Some of the areas where avoidable costs can be calculated include:
1. Preventive and predictive maintenance (avoids function loss breakdowns)
2. Repair and replacement maintenance (correcting function loss breakdowns)
3. Improving and restoring maintenance (reducing function reduction breakdowns
4. Compliance with governmental regulations (environmental, safety, and hazardous materials)
5. Cosmetic appearance (for operators and customers)

Preventive and Predictive Maintenance

The preventive and predictive maintenance program requires the expenditure of resources with the calculated probability of avoiding function loss breakdowns. This expenditure must be weighed against the probable cost of the breakdown in order to keep the program operating in a cost effective manner.

Repair and Replacement Maintenance

Repair and replacement are usually in response to a problem that was noted during the inspection and monitoring portion of the preventive and predictive maintenance program. Once a potential problem is noted, repair or replacement can be planned in a way that causes minimal interruptions to the operations department, yet still prevents or corrects a breakdown.

TOTAL ECONOMIC MAINTENANCE

Improving and Restoring Maintenance

Improving or restoring equipment after a function loss breakdown usually involves cleaning or replacing a component that is on a trend of decreasing efficiency. These tasks include cleaning pumps with decreasing efficiency or restoring the full operational capacity to production equipment. This area of maintenance improvement is generally overlooked.

Compliance with Governmental Regulations

Compliance with governmental regulations can also be a function of maintenance. For example, one municipality was fined over $1 million for an environmental violation that was ultimately traced to a pump that was improperly maintained. Fines are a small part of the total picture when compared to environmental damage, employees' health, or damage resulting from improper handling of hazardous materials are considered.

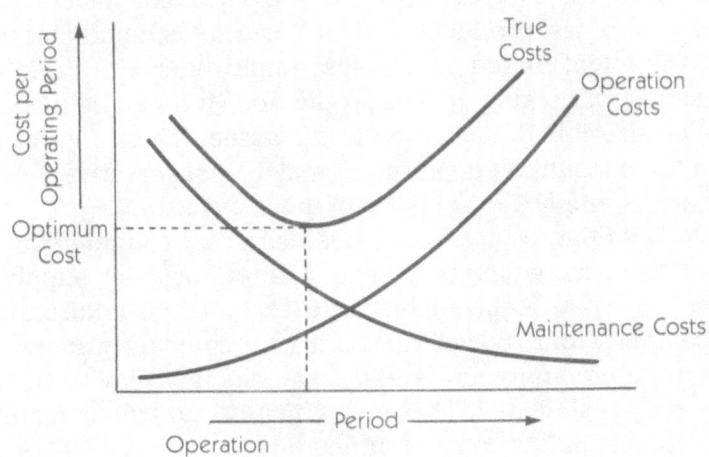

Figure 11-1 Maintenance timing decisions can be made by establishing the lowest cost ratio between maintenance and operations.

158   TOTAL PRODUCTIVE MAINTENANCE

Cosmetic Appearance

In the TPM mindset, the environment in which employees work, the impression that the operation has on customers, and the public's opinion of the company may not easily translate into dollars, but can be factors. For example, studies estimate that over 20% of all manufacturing costs are related to quality problems. Understanding how maintenance affects quality costs is important to being able to quantify any costs in this area.

Figure 11-1 illustrates this costs vs. benefit discussion. The figure shows that the decision for maintenance would be made based not on what is best for the operations group, nor what is best for the maintenance group, but on what is the lowest combined cost. This cost is the effective bottom line for the company. This type of decision is one that companies must make if they are to optimize their resources.

Calculating Costs

Collecting Information

How does a company collecting the information required to perform some of these analysis? It starts by assigning a cost to downtime. The accounting department can help determine the costs of an hour or a shift of lost production when a piece of equipment is down. These costs might include lost sales, employee salaries and overhead, the cost of making up lost production (if it can be made up), and any measurable depreciation to the assets. These figures coming from the accounting department may be conservative, but they will not be disputed by other parts of the organization.

With these figures agreed to, it is necessary to understand the maintenance costs, which may include labor, material, supply, and miscellaneous costs incurred due to the repair or the failure. Having both repair and failure costs is necessary for comparing an overhaul to a run-to-failure approach. Additional costs that may be incurred should also be calculated. These may include hazardous materials, EPA, OSHA, and safety considerations.

Identifying Problems

Once the total cost is understood, some interesting problems can be considered. The following problems address scenarios in

three of the areas of concern to companies. These are:
1. Preventive maintenance frequencies
2. Critical spares analysis
3. Normal maintenance parts and supplies

Preventive Maintenance Frequencies

When examining PM frequencies, a sample common to most plants or facilities will be selected: a centrifugal pump. Whether this pump is pumping a product or moving cooling water, it has a value for its service. Setting a price on this value give a reference from which to start. Suppose the value is $100 per hour. Managers can

---
BASIC MAINTENANCE COST CALCULATION

Repair Cost = $1,500

If the pump was serviced once every 100 hours, the cost would be:

$$\frac{1500}{100} = \$15.00/hr$$

If the pump was serviced once every 500 hours, the cost would be:

$$\frac{1500}{500} = \$3.00/hr$$

The table version would be:

Service Frequency (Hours)	Maintenance Cost (Dollars)
100	15.00
500	3.00
1000	1.50
1500	1.00
2000	0.75
2500	0.60
3000	0.50
3500	0.43
4000	0.38

Figure 11-2 A method for calculating the maintenance cost of a pump

160   TOTAL PRODUCTIVE MAINTENANCE

now solve the following problem. Assume it costs $1,500 for parts and labor to overhaul the pump. There is no downtime cost because a standby pump is available. A measurement of the pump's performance indicates that after 4,000 hours of operation, it loses 5% of its capacity. (Assume the drop is linear and continues to be so throughout the life of the pump.) When is it cost effective to remove the pump from service and clean the rotor?

This problem can be solved by calculating the amount of maintenance cost and the lost performance cost per hour. The two measures are combined to give the lowest total cost. The techniques for performing this analysis are illustrated in Figures 11-2, 11-3, 11-4, and 11-5. In Figure 11-2 the maintenance cost is calculated. This particular example highlights the mistake of considering only maintenance costs. If only these costs are considered, it would be advisable to delay servicing the pump for as long as possible.

LOST PERFORMANCE CALCULATION

If the performance loss is linear, then at 4000 hours of operation, the loss is 5% and the value is $100.00, then the value of the loss is:

$$0.05 \times \$100.00/hr = \$5.00/hr$$

Therefore, at 4000 hours of operation, the pump is producing only a value of $95.00, or it is losing $5.00 per hour.

The table version would be:

Time Since Last Service (Hours)	Lost Performance Cost (Dollars)
100	0.13
500	0.63
1000	1.25
1500	1.88
2000	2.50
2500	3.12
3000	3.74
3500	4.36
4000	5.00

Figure 11-3 A simple method for calculating the performance falloff of a pump.

TOTAL ECONOMIC MAINTENANCE   161

Figure 11-3 indicates that when you delay the service, the amount of lost production cost increases linearly. The difference between Figures 11-3 and 11-4 is that the latter takes into consideration that performance falloff is triangular and not the total area volume of the rectangle.

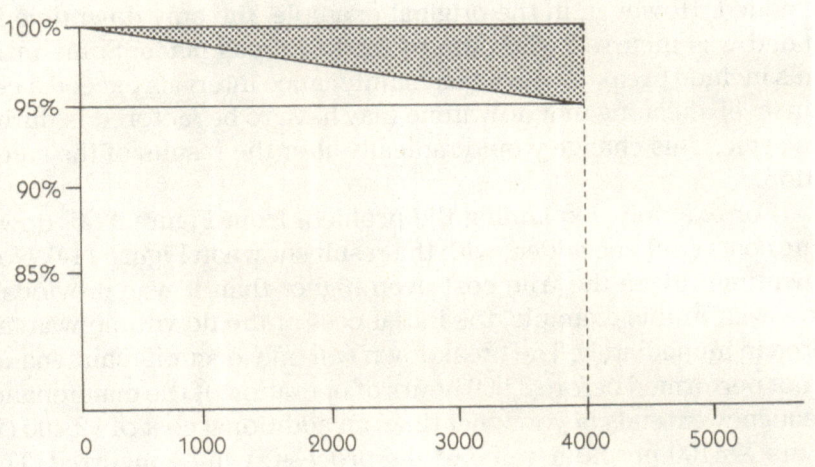

The total loss is not the entire rectangle, but only ½ of its area. So the loss would only be ½ of the calculated amount. The revised version of the table would be:

Time Since Last Service (Hours)	Lost Performance Cost (Dollars)
100	0.065
500	0.31
1000	0.63
1500	0.94
2000	1.25
2500	1.56
3000	1.87
3500	2.18
4000	2.50

Figure 11-4 Calculating the actual performance falloff of a pump.

162 TOTAL PRODUCTIVE MAINTENANCE

The problem's summary is illustrated in Figure 11-5, which plots the falling maintenance cost against the increasing cost of the lost performance. These two factors added together give the total of the true costs. The decision can then be made on lowest true cost. In this problem, the lowest true cost indicates that maintenance action should be performed every 1,500 hours of actual run time.

The problem is usually more complex; this problem started with no penalty cost for the downtime required for maintenance action. If downtime has additional costs, then another column would have to be added. However, in the original example, the only downtime included was incurred when the repair itself was made. Some problems include breakdowns when maintenance intervals exceed a certain level, meaning that downtime may have to be factored in during the cycle. This change would radically alter the results of the calculation.

For example, expanding the problem from Figure 11-5, downtime costs could be added, with the result shown in Figure 11-6. Now downtime drives the true cost even higher than it was previously. However, in this example, the initial cost of the downtime was factored in immediately. The breakdown will only occur if maintenance is not performed before 3,000 hours of operation. If the maintenance frequency extends beyond that time, an additional cost of $2,400 (24 hours X $100 per hour value of the process) will be incurred. This

TRUE TOTAL COST CALCULATION

The true total cost can now be established for this task. Combining the tables results in:

Time Since Last Service (Hours)	Maintenance Cost (Dollars)	Lost Performance Cost (Dollars)	True Total Costs (Dollars)
100	15.00	0.065	15.065
500	3.00	0.31	3.31
1000	1.50	0.63	2.13
1500	1.00	0.94	1.94
2000	0.75	1.25	2.00
2500	0.60	1.56	2.16
3000	0.50	1.87	2.37
3500	0.43	2.18	2.61
4000	0.38	2.50	2.88

Figure 11-5 A chart can be used to calculate the lowest true cost. This is the optimum opportunity for maintenance action.

FACTORING IN BREAKDOWN COSTS

If a breakdown occurs and there is no spare, then the service also results in a cost penalty to the Operations Group.

— For planned service—8 hours of downtime
— If a breakdown occurs—24 hours of downtime

A breakdown will occur every 3000 hours of operation, based on repair records.

The table shows:

Time Since Last Service (Hours)	Maintenance Cost (Dollars)	Lost Performance Cost (Dollars)	Downtime Cost (Dollars)	True Total Costs (Dollars)
100	15.00	0.065	8.00	23.07
500	3.00	0.31	1.60	4.91
1000	1.50	0.63	0.80	2.93
1500	1.00	0.94	0.53	2.47
2000	0.75	1.25	0.40	2.40
2500	0.60	1.56	0.32	2.48
3000	0.50	1.87	1.07	3.44
3500	0.43	2.18	0.91	3.52
4000	0.38	2.50	0.80	3.68

Figure 11-6 Adding breakdown costs to the calculation provides a more realistic estimate of true total cost.

calculation removes any doubt that the lowest total cost would be around 2,000 hours of operation, but definitely before 3,000 hours of operation.

The ability to apply statistical techniques can go far beyond this simple example. Consider the ability to perform this type of costing for each subcomponent of a large mechanical drive; the ability to determine the lowest life-cycle costing for complex equipment; the ability to determine the amount of resources to be spent on redesigning and retrofitting engineering projects, based on anticipated return on the investment.

However, while this model is a straightforward mathematical one, why are so few companies doing this type of calculations? The problem is very few of them have any reliable data with which to work. This is why the program outline for the American approach to TPM places this step after good work order controls are implemented. The data needed for these calculations must come from the work order system. Without this reliable data, companies will be back to guessing when maintenance activities should take place and not optimizing their resources.

Critical Spare Analysis

A second example of statistical controls for maintenance examines critical spares. Spare parts are an albatross for maintenance managers. The operations and upper management see critical spares (which are typically slow moving) as unnecessary or wasted because they never seem to turn over. The same techniques that was applied to the equipment can be applied to the spares. For example, consider what costs are factored into a stocking decision:

> How many are used per year or Mean Time Between Failures?
> What is the operating requirements for the equipment per year?
> What is the cost of downtime if the part is unavailable?
> What does the item cost?
> What is the annual holding cost for the item?
> How often can the item be repaired?
> What is the lead time to get a replacement?
> What is the lead time to repair the item?

After all the costs related to the stocking decision are identified, a logical decision can be made concerning the stocking level. For example, the difference of keeping no spares and keeping one spare could be substantial when resulting downtime is factored into the calculations. Critical spares typically have long lead times. If no spares are kept, the resulting downtime may be weeks of lost production. Costs can quickly add up to tens of thousands of dollars. Consider the following example.

A gearcase costs $24,500 and is a spare on one equipment unit. When it is ordered, it takes one week to receive the spare. The holding cost is 30% of the price of the gearcase. While the gearcase is down, the downtime cost is $1,000 per hour. The actual time to replace a gearcase is 40 hours and the labor cost is $1,000. Currently four spares are stocked; annual usage for the past three years shows

TOTAL ECONOMIC MAINTENANCE   165

Project Results:

	Total Cost	Impact Costs	Purchasing & Holding Costs
0	$209,000.00	$209,000.00	$.00
1	$72,900.00	$41,000.00	$31,900.00
2	$104,800.00	$41,000.00	$63,800.00
3	$136,700.00	$41,000.00	$95,700.00
4	$168,600.00	$41,000.00	$127,600.00
5	$200,500.00	$41,000.00	$159,500.00
6	$232,400.00	$41,000.00	$191,400.00
7	$264,300.00	$41,000.00	$223,300.00
8	$296,200.00	$41,000.00	$255,200.00
9	$328,100.00	$41,000.00	$287,100.00
10	$360,000.00	$41,000.00	$319,000.00

"Figures supplied courtesy of GenesisSolutions – www.genesissolutions.com"

Figure 11-7

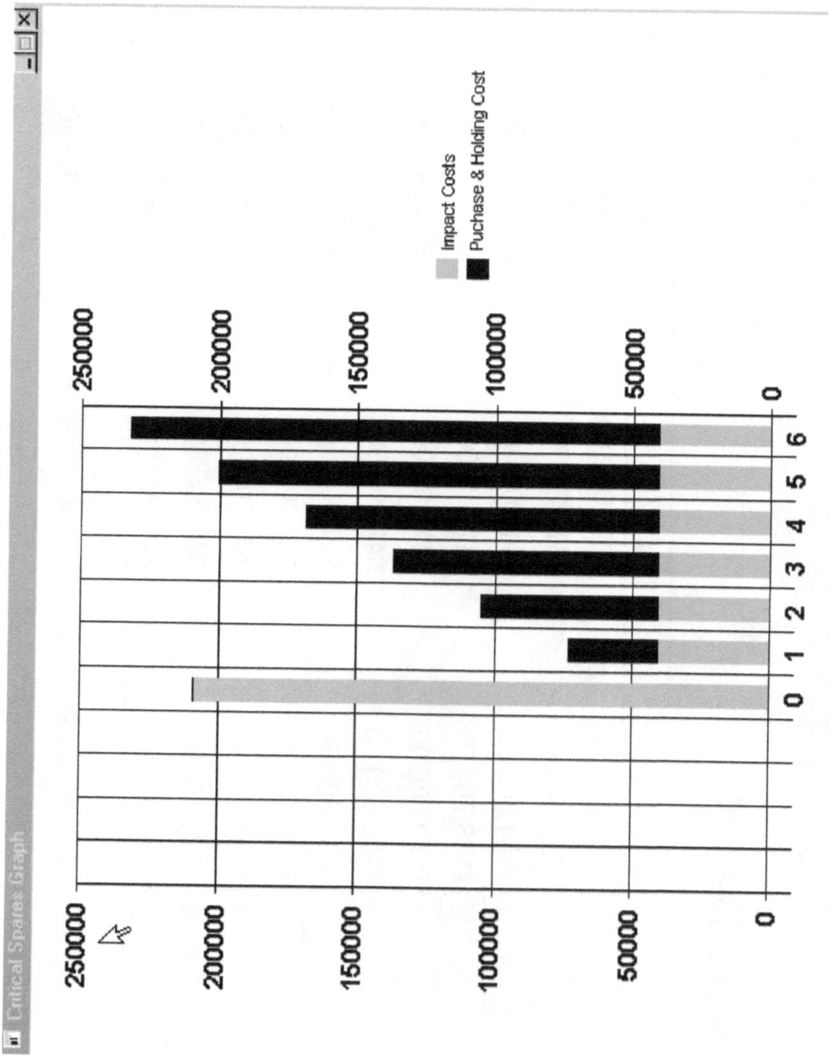

166 TOTAL PRODUCTIVE MAINTENANCE

Figure 11-8

TOTAL ECONOMIC MAINTENANCE

that one gearcase was changed each year. Based on this information how many spares should be stocked and why?

Keeping in mind the total cost curve, you must plot the cost of investment (stocking the part) versus the cost of downtime during the failure. This can be performed in a table format (see Figure 11-7) or in the form of a graph (see Figure 11-8).

When the comparison is made, it is clearly seen that keeping one spare is the optimum financial decision for the company. The difference between keeping one spare (optimum) and keeping four spares (current policy) is $95,700. This difference represents considerable savings for just one component. Most companies have hundreds of these types of spares. Consider the savings possibilities for optimizing critical spares across all components. Also consider the saving potential for an organization that reduces inventory without factoring in the cost of downtime. The difference between keeping none and keeping one is $136,100, due mainly to the resulting downtime and lost production. The savings resulting from keeping one spare has a great impact on the company's overall profits. Only when companies take this financial approach to stocking maintenance spares stocking will inventories truly be optimized.

Normal Maintenance Parts and Supplies

The same calculations could be applied to normal stores items. If the same information is kept, statistical formulas can be used for calculations such as service levels, actual cost per item, and number of projected turns per year. These can provide a manager with the tools necessary to make accurate and cost effective management decisions.

The limitation to all of the calculations is the need for accurate data. Maintenance organizations must have the discipline to collect accurate data before any of the techniques in this chapter can be used. Statistical techniques undertaken with poor or inaccurate data will provide just as an inaccurate answer as guessing. A good, disciplined approach to the work order system will enable may of the techniques mentioned in this chapter to be useful to any company attempting to optimize its maintenance expenditures.

CHAPTER 12

TEAM-BASED MAINTENANCE

Team-based maintenance is what companies envision when they hear the term TPM. If the plan presented in this book has been followed, the next step is to evaluate the types of organizational practices presently used.

A Progress Check

By this point, the organization should be experiencing World Class results. The organization is dedicated to using a disciplined approach to the work order system. All information is recorded completely and accurately through the work order system. The effective use of the work order system allows for accurate scheduling practices. Underlying the entire structure of the maintenance organization is the attitude of maintenance prevention. The preventive and predictive maintenance programs are in optimum condition. The costs related to PM/PDM are calculated in the scope of the lowest total cost to the company.

These changes show employees that the maintenance improvement program is not just another program of the month. Management's dedication to this program has been proven by the expenditure of the resources needed to accomplish the improvements to date. During the time that has passed to bring the organization to this stage, the employees' attitude has become one of cooperation and understanding.

Transferring Tasks

If, at this time, management begins to transfer basic maintenance tasks such as routine inspections and service, the operations and maintenance groups will cooperate. The plans for this transfer must be presented to the employees in the context of total costs. The benefits and results must be quantified for the employees, not presented in vague terms such as "It is the best for everyone if we do this." Clear, distinct financial data should be developed and presented. Management should make clear to the employees that the program's objective is not to eliminate any maintenance jobs, but instead to raise the level of maintenance on the equipment. If the operators inspect, clean, and routinely service their equipment, they will contribute to increased equipment effectiveness.

The new relationship between operations and maintenance can be compared to the relationship between paramedics and doctors. The paramedics are the first on the scene, applying first aid to the patient. If there is still a crisis with the patient, then the doctor (maintenance technicians) needs to see the patient as quickly as possible. If the emergency passes, the patient should still be scheduled for an appointment so the doctor can perform any routine or additional maintenance tasks.

What is the biggest benefit to the company of transferring tasks? In addition to seeing the equipment in a constantly improved condition, the typical minor stoppages are even further reduced. As a result, the company's profit margin will continue to increase, making it even more competitive than before.

How is operator-based maintenance implemented? What are the program's limitations? What are the pitfalls that may be encountered? These questions will be addressed in the remainder of this chapter.

Operator-Based Maintenance

How is operator-based maintenance implemented? The implementation steps that most companies can use are as follows:

1. Identifying the tasks to be transferred
2. Training the operators for the tasks
3. Monitoring the program

Identifying the Tasks To Be Transferred

In order to identify the tasks to be transferred, the company must understand what is being accomplished. The work itself must be transferred in blocks of ten minutes. If the operators are required to spend any more time performing maintenance tasks during a shift, their core jobs will suffer. Some companies allow 10 minutes per shift and one 30-minute period some time during the week. If it allots any more time to maintenance tasks, the company will experience a proportionate decrease in production levels.

The tasks that are to be transferred are routine inspections, adjustments, and small services. These include the inspections for looseness, unusual wear, contamination, and leaks. The objective is not to have the operators fix all of the problems that are discovered, but instead to fix what they can and alert maintenance about the need for repair for the remainder. Whatever maintenance tasks cannot be completed by the operators in the allotted time must be performed on a planned and scheduled basis by the maintenance department. The adjustments include tightening any loose parts to prevent vibration, insuring the proper adjustment of all parts of the equipment the operator contacts, and lubricating components as required.

How are these tasks identified? They are selected from the existing preventive maintenance program. If the present preventive maintenance tasks are properly detailed (see Chapter 6), then selecting the tasks to be transferred becomes simple. This process is one of the main reasons why developing the preventive maintenance program (see Chapter 5) is so important. Many of the present PM tasks, once adjusted into smaller blocks of time, can easily be transferred to the operators without having to create a new PM program. Any tasks that are too complex or unsafe for the operators to perform should remain part of the maintenance program.

Training the Operators

Some companies think that their operators should already know how to perform the maintenance tasks relating to their equipment. This myth must be dispelled quickly. Just as all maintenance technicians require training to do their tasks, the operators need training as well. It takes only one missed problem on an inspection, one incorrect adjustment, or one missed lubrication point for a breakdown to occur.

Over time, if breakdowns continue to occur, the operators and the maintenance technicians will become increasingly frustrated, and the entire program will rapidly deteriorate. It would be extraordinarily wasteful to bring the organization this far, then lose all progress because of failing to invest in operator training.

Safety is another important reason for providing the operators with proper training. Otherwise, accidents will occur. This concern will provide the skeptics more problems to point at within the program. As a group, operators will only have some many accidents before they quit performing the tasks.

Given the importance of training, where do you get the material to train the operators? If the plan has been followed, detailed maintenance training materials will have already been created (see Chapter 10). These materials can be adapted to the tasks that the operators now perform, providing sufficient training to assure competency. The maintenance technicians who work on the team with the operators can provide the actual training which, like the maintenance training, should include a mix of written instructions and actual monitored performance of the task. This process also helps to provide a bonding between the operators and the maintenance technicians. Some companies even provide a certificate of completion each time an operator masters an assigned task.

The training must be complete to ensure that the operators learn enough to set their own standards for:

Cleaning	Adjustments
Lubrication	Operational Parameters
Inspection	Housekeeping
Set ups	

Without adequate training, the continued reliance on the maintenance department will negate any advantages of going to operator-based maintenance. Training the technicians is equally important. After all, if they are training the operators, then they must be technically proficient to perform the tasks listed above. Unfortunately in most companies today, very few technicians know enough to set maintenance standards in these areas.

Monitoring the Program

Monitoring the program is a matter of measuring the program's effectiveness. The equipment effectiveness is the ultimate measure of the program. If it is not improved, then the tasks being performed by the operators need to be examined. Are they adequate? Are they eliminating or preventing problems? If not, then appropriate steps need to be taken. The "continual and rapid improvement" which is the world class theme must be applied to program monitoring. The key to almost all problems in operator-based maintenance is training, training, and more training.

The transfer of selected maintenance tasks to operations should in no way signal a defeat to maintenance. The next step for the maintenance organization is to begin advanced training for its technicians. This training allows the in-house technicians to take over the more advanced role of being a direct interface to engineering, outside vendors, and technical consultants. The transfer of the maintenance tasks to operations is limited. Estimates show that approximately 10% to 40% of the maintenance tasks mentioned previously can be transferred to operations. The time freed up by this transfer should then be applied to learning to maintain and repair new technology, allowing the organization to continue to grow and become competitive.

The Manager's Role

Because the relationship between operations and maintenance change under the TPM banner, the manager's role must also change. When the operators and technicians work together, they often form problem-solving teams. These teams may be assembled to solve specific problems, or they may be used to monitor certain operational parameters. They solve problems and generate improvements. By using teams, managers can delegate control and decision making to the lowest possible level within the organization. This process, however, dramatically changes the manager's role from monitor to coach. In fact, the traditional management role evolves into a new World Class role, which can be divided into the following four main areas:

1. Recognize the importance of the work
2. Help set and achieve the goals
3. Take action on the suggestions
4. Reward the employees' efforts

Recognize the Importance of the Work

If the employees do not believe that their work is important, they will put little or no effort into the program. The supervisor's attitude will determine the importance the team puts into the improvement effort. Remember: If you don't think their efforts are important, they will not either.

Help Set and Achieve the Goals

Supervisors use their own knowledge to steer the team to work on problems that, when solved, can support the corporate goals. Without this guidance, the team is working as if it were running a race without a lane or finish line. By providing direction and clarifying goals and proposed time frames for the improvement projects, the supervisors help set the finish line. People work better when they have clear goals and objectives. The balancing act for supervisors is to steer the group to this information, but not to dictate it to them.

Take Action on the Suggestions

Supervisors next have the task of acting on any suggestions that the team develops. Supervisors must help the teams by seeing that suggestions are properly written, perhaps providing teams with the cost information or engineering data that is required to substantiate the suggestions. They may even need to present the suggestion to upper management, if the team feels insecure in doing so. Once the plan has been presented, supervisors can follow up to insure implementation. This follow up may be in the form of helping the teams overcome problems that develop.

Reward the Employees' Efforts

Rewards, which help to satisfy an individual's desire for recognition, generally take three forms:
- Monetary rewards
- Plaques or certificates
- Formal notice presentations

Supervisors must know both the make up of the team and the type of reward to use for each suggestion. Recognition will help motivate the team members to achieve even more. It also helps to stimulate the creativity of other teams as well. Positive competition can greatly benefit the corporation.

As can be seen from the discussion above, the supervisor's role changes from manager to advisor or coach. This transition takes training and patience on the part of the supervisor. In turn, those who manage supervisors should be able to help them make the changes required to manage teams in the new environment.

Steps to Autonomous Groups

Autonomous groups are teams of employees who work together, independent of direct supervisory control. The supervisors still function as coaches, as previously described. However, autonomous groups continue to grow to maturity. They progress through four distinct phases:
1. Self Development
2. Improvement Activities
3. Problem Solving
4. Autonomy

Self Development

In the self development phase, the group begins to see the larger picture from the corporate position. They start seeing their role within the company and the contributions that they can make. They must develop to the point where they learn new techniques and gather new information. They become proficient with understanding and applying the new information and techniques to their jobs. As they begin to mature, both as individuals and as a group, they begin to understand the importance of each team member and the contributions that each member can make to the group. They learn to be comfortable working within a small group.

Improvement Activities

Improvement activities begin when the group develops suggestions that are proposed and implemented. At this stage, supervisors are still choosing projects or areas that need improvement. When the suggestions are implemented, they produce results. The results show improvement, which in turn produces a sense of accomplishment in the team members. This sense of accomplishment allows the teams to take pride in their work, producing an even more intense desire to make additional contributions.

Problem Solving

The problem solving step is a selection process that allows the group to prioritize the list of problems, focusing on those that are of particular importance to their departments or specific job duties. The group has the independence of no longer being told which problem to work on. This step sets the stage for the next development of autonomy. In this problem solving stage, supervisors are involved on an as needed basis. They are no longer needed to select the specific problems for the group. This phase is often difficult for the supervisors as their roles begin to change dramatically.

Autonomy

The last phase is the autonomous group. In this step, the employees have matured to the point of requiring very little, if any direction. Teams are now able to select goals that are consistent with corporate policies and are fully capable of managing themselves. This form of employee empowerment produces dramatic results within an organization. With the decision making pushed down to the lowest possible level, the theme of continuous and rapid improvement becomes a practical way of life, not just a concept. The organization that reaches this stage has truly become a World Class company.

Summary

This chapter has examined the concept of team-based maintenance. Keep in mind, however, that to date few true success stories exist. Some organizations start well, but due to management changes, economic constraints, or rapid growth, begin to deviate from the process that made them successful. In turn, the TPM work culture begins to devolve back to more of a traditional focus, ultimately ending the team-based environments. Only when companies can stay focused on the principles outlined in this chapter will they ever have a sustainable TPM work culture.

CHAPTER 13

PERFORMANCE INDICATORS FOR TPM

This chapter looks at factors that are important for measuring the results of a maintenance improvement program. The first challenge is to select the maintenance indicators that should be used. Consultants will try to sell their clients on hundreds of different indicators. Although there are so many indicators, the focus should be on the financial bottom line. The total measure of maintenance performance is the percent of sales dollars spent on maintenance. In the case of facilities, that measure is the maintenance cost per square foot maintained. This total measure is critical to meeting one of the goals of world class manufacturing or service: providing all products and services at the lowest cost.

Performance Indicators

If a competitor has a maintenance cost equal to 4% of the sales dollar and your maintenance cost is 7%, then the competitor enjoys a 3% advantage. That competitor can use that advantage to increase profit, reduce price, fund new capital investment, or expand research and development. That difference puts your company at a definite disadvantage.

Three main categories of indicators can be used to measure performance:

    Financial
    Effectiveness
    Performance

PERFORMANCE INDICATORS FOR TPM

The relationship among the various types of indicators is pictured in Figure 13-1. Each level highlights an indicator. The cause of any problem and the subsequent solution is provided by examining the indicators in the next lower level. By the time any problem reaches the bottom level either the supervisor, planner, or craft technicians should be able to provide a solution.

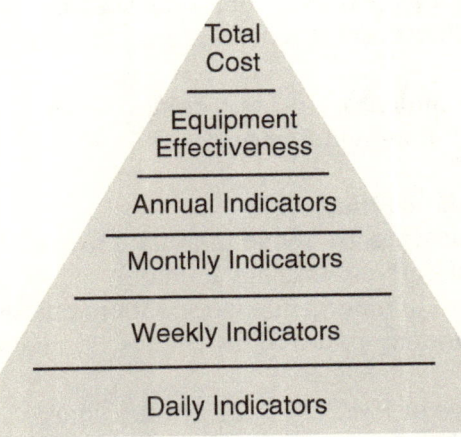

Figure 13-1 The maintenance performance measurement pyramid

Financial Indicators

Financial indicators were discussed in Chapter 11. In this area, we must move away from the traditional method, the mindset that 'We have done it this way for years, why change?" Only by examining the total maintenance cost can we become competitive. The top of the pyramid include techniques that only the most progressive companies are utilizing.

Effectiveness Indicators

The effectiveness of the maintenance organization (Level 2) should always be measured by the overall equipment effectiveness formula (see Chapter 3). This formula highlights areas that could be misinterpreted to indicate maintenance problems. However, when the formulas that in turn make up the overall equipment effectiveness formula are examined, the true source of the problem is uncovered.

Performance Indicators
Annual Performance Indicators
Some of the annual indicators that can be used to trend maintenance performance include:
1. Maintenance cost as a percentage of estimated replacement value
2. Maintenance cost as a percentage of asset purchase price
3. Maintenance costs as a percentage of total sales
4. Maintenance costs as a percentage of square foot maintained

While these indicators do not comprise all that might be selected, they are common indicators used for long-range trending.

Monthly Performance Indicators
Monthly indicators are those that need monitoring over a time period to trend effectiveness. They include:
1. Percentage of emergency work: hours charged to emergency work compared to the total maintenance hours expended.
2. Percentage of PM/PdM work: hours charged to PM/PdM compared to the total maintenance hours expended
3. Percentage of downtime caused by breakdowns: hours of downtime caused by breakdowns compared to the total number of downtime hours
4. Percentage of overhead: total dollars spent on maintenance overhead compared to the total maintenance labor dollars expended.
5. Percentage of maintenance labor and materials costs: the total labor cost compared to the total material cost.
6. Craft Backlog: the amount of work ready to schedule in the craft backlog compared to the average amount of scheduled work completed each week.

Weekly Performance Indicators
Weekly maintenance performance indicators measure actual maintenance performance. and are on the next-to-bottom level on the pyramid. Some common indicators are:
1. Schedule compliance: number of work orders scheduled to be completed compared to the actual number of work orders

PERFORMANCE INDICATORS FOR TPM

completed
2. Supervisor performance report: compares estimated labor and material requirements for scheduled jobs to the actual labor and material used
3. Planner performance report: compares what was planned for the week versus what was actually completed
4. Work order costs report: compares what was actually planned versus what costs were actually charged for the weeks work by individual work orders.
5. Technicians' time: the percentage of time spent in emergency, preventive maintenance, normal scheduled, and out age work assignments.

Daily Performance Indicators

Daily indicators, which measure the performance of maintenance activities on a day-to-day basis, make up the bottom level on the pyramid. These measures supply the information on which the rest of the pyramid is built. They include:
1. Work completed: compares the actual versus the estimate for each work order completed during the previous day
2. Emergency work requests: lists all breakdown and critical work completed the day before
3. Preventive maintenance work due and overdue: lists all PM work that is due and overdue as of the current date
4. Stock replenishment list: lists all stock items that have fallen below their minimum and need reordering in the last 24 hours
5. Technicians time: lists each work order and the time spent on it by each technician for the last 24 hours

This list of performance indicators is not complete, but rather a list of suggested or common indicators. Some organizations will use additional or different indicators. It is important to select upwardly compatible performance indicators so that the top of the pyramid has the information necessary for making good decisions based on the data developed. (For further information on performance indicators, see Performance Indicators for Managing Maintenance, Industrial Press, 1998.)

What To Do with the Data

A common mistake organizations often make is to neglect benchmarking the status of the maintenance organization before starting a TPM program. The steps outlined in this book include that vital step; therefore, you should have a frame of reference for this entire project. Benchmarking the results and then reporting on them helps keep management support for the program.

At least once every six months, provide upper management with a report that updates the status of the project. Any update should summarize the costs versus the benefits. Realistically, money will be spent at the start of the program before benefits are achieved. However, as long as such expenses are according to the plan, management should have no problem. Instead, problems occur when the results deviate from the plan. The data you have been gathering can be used to explain any deviation as well as corrective steps being taken before the problem becomes critical.

It has often been said that "what gets measured gets managed." If the measurement techniques described thus far are used, the effort required to convince management that maintenance is an important contributor to the corporate profit picture will be simplified.

CHAPTER 14
THE FUTURE OF TPM

Where will TPM go in the future? The answer lies in three basic areas:
1. Management Acceptance
2. Technology
3. Management Techniques

Management Acceptance

The key to management acceptance is educating management about the value of TPM. Understanding how to assign a dollar value to the Overall Equipment Effectiveness calculation was highlighted in Chapter 3. Some of the techniques from Chapter 11 about Total Economic Maintenance will also help. There are also other opportunities, some of which can be learned from studying the history of quality programs.

There are amazing similarities between the obstacles that TQC programs had to overcome and those facing maintenance organizations. Consider the works or Crosby, Deming, Juran, and others. Learn how they convinced management that it was time to take action. Also study the methodologies they used to implement their ideas because they have major application to maintenance programs.

Consider Figure 14–1. The pyramid in part A is typical of those American industries that tried to implement all of the various World Class programs, but neglected to provide a strong foundation (usu-

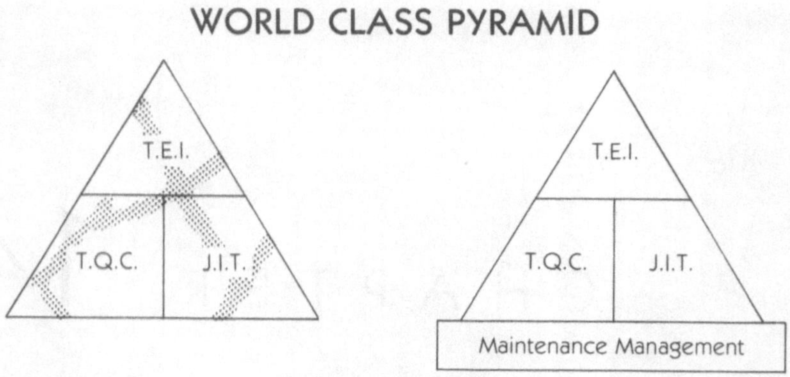

WORLD CLASS PYRAMID

Figure 14-1 Maintenance management provides a sound foundation for industry.

ally their assets). Over a period of time the programs begin to crack and crumble, and the shifting sands of the market and competition take their toll. In part B, a strong foundation of maintenance basics, coupled with world class programs, presents a solid pyramid without any cracks. Because the assets support the goals and objectives of the other programs, the company can remain competitive, no matter what challenges are thrown against it.

Only by convincing management of the true value of maintenance, and by extension asset management, can any improvement ever be made. Without convincing management, the funding and disciplines will never be provided.

Technology

Some people claim that in the area of technology, there is nothing new under the sun. The last decade has seen continual refining of some technologies, but no new ones have been introduced. Predictive maintenance will continue to be emphasized, but new technologies will not be introduced to revolutionize this market.

Perhaps the greatest technological change has been the use of expert systems to help maintenance technologist be more effective in troubleshooting equipment. This technology, coupled with the real time monitoring of data through PLCs (programmable logic controllers) and DCSs (distributed control systems), will enable the technicians to troubleshoot more accurately than ever before. The question will still revolve around the funding necessary to train the technologists to take advantage of these tools.

Management Techniques

The final area to be considered is that of management techniques. Evaluate your maintenance supervision and their skills, then ask whether they are capable of managing in a World Class environment. Many employees who are currently in supervisory and management positions in maintenance are 45 years old or older; they are beginning to think about retirement. They are more hesitant to begin implementation of a new business process (like TPM) that would require a huge amount of energy just before retirement. Some even have the attitude that "We have always done it this way." The techniques and skills required to manage in a TPM environment are entirely different. Will they be able to adapt?

A second problem is the insufficient preparation of the new maintenance managers who are replacing the current maintenance managers as they retire. Many are coming from outside the maintenance field. It takes time to educate them about maintenance management because the discipline is not like any other management discipline in a company. Will the new managers be able to learn not only how to manage maintenance, but also to understand it well enough to continue leading it to progressively higher and higher levels of efficiency and effectiveness? These are indeed challenges facing the maintenance business today.

Summary

Maintenance is not hard to do; it is only hard to sell. The concepts and techniques that are hailed as new and innovative are really only variations of older techniques that never became popular when companies did not have to be competitive. The question now facing managers in companies today is whether they want to change to be competitive or not change and become extinct.

CHAPTER 15
MAINTAINING THE TPM VISION

This chapter summarizes the textbook and highlights thoughts on how to maintain the TPM vision for your company.

At the start of this chapter, it is appropriate to ask the reader to consider the following question: If you could describe the main goal or objective for TPM in one statement, what would it be? Do not answer this question yet. From reviewing this chapter, the answer will be clear.

General Review

When considering Chapter 3's discussion about OEE, it should be clear that the only thing more expensive than developing a TPM strategy is to do something else instead. Unless the company's assets are capable of supporting the improvement program, any other World Class improvement program will require more time and resources than it should to implement. If a company's assets perform at the level they were designed to perform, then initiatives such as Just in Time, Total Quality, and Lean Manufacturing will succeed. Otherwise, they will not; it is just that simple.

In looking over the entire book, some might be tempted to question how many companies are really implementing TPM, as if to say that because few others are using it, they should not have to use it either. However, that attitude is defeatist. The correct attitude is one that observes "I see the value of TPM; but our organization is not quite ready yet. What are some first steps we can take to start our company in the direction of TPM?" The value of TPM is itself not open to dispute. The real issue is whether your company can be educated to see that value and then make steps to develop and implement a TPM strategy.

Chapter 1 introduced a TPM decision tree that provides a framework of steps for developing a TPM strategy. Any organization that is beginning a TPM program, restarting a stalled TPM initiative, or trying to improve an existing TPM strategy, should examine the decision tree for areas to improve. Some organizations will have to start with the basics, whereas others will build on the basics already in place, continuing to advance their TPM maturity. The decision tree should be used periodically to evaluate any TPM program, insuring that it is always moving ahead and making the company more competitive.

Remember too that TPM should always be supported from the top, but driven from the bottom. Thus, TPM should always have senior management support. At the same time, TPM should be driven or directed by the personnel actually operating and maintaining the equipment. They are the ones best suited to develop ideas for making the equipment more productive. The OEE should always be the measure used to track improvements.

TPM Failures

When post mortems are performed on failed TPM strategies, four basic reasons for failure emerge. They are:
1. Insufficient understanding of the basic concepts of TPM
2. Lack of qualified personnel to teach TPM
3. Lack of flexibility in implementing TPM
4, Lack of a complete vision of TPM

Insufficient Understanding

Recall from Chapter 1 the five basic goals or pillars of TPM:
1. Improving equipment effectiveness
2. Improving maintenance efficiency and effectiveness
3. Early equipment management and maintenance prevention
4. Training to improve the skills of all people involved
5. Involving operators (occupants) in routine maintenance

If only one goal is the focus for TPM, the program becomes one dimensional. For example, if a company focuses only on operator involvement, it misses input from maintenance; without an effective maintenance organization, the TPM initiative will struggle and fail.

Lack of Qualified Personnel

The lack of qualified personnel to teach TPM relates to the neglect of developing internal champions of TPM. If only one person has the vision, it is unlikely that person will be able to communicate it to all levels of the company. Others will have to become champions as well. When TPM champions are active at all levels within the organization, they can influence their peers. An understanding and enthusiasm for what TPM could mean for the entire company will become uniform. This understanding will accelerate TPM activities.

Lack of Flexibility

The lack of flexibility in implementing TPM relates to an important point made in Chapter 1: There is no cookbook approach to TPM. The program must be introduced in a way that fits the individual organization. Issues such as skill level of the workforce, union-management relations, business conditions, and type of industry must all be addressed. Without addressing these issues for each company, TPM will have virtually no chance at succeeding.

Lack of a Complete Vision

The lack of a complete vision for TPM further highlights the one-dimensional approach that many companies have to TPM.

For example, some organizations start by cleaning their equipment. They put a lot of resources, both labor and material, in cleaning their equipment and organizing the workplace, but do not incorporate these resources into the context of a total TPM strategy.

On the day that upper management and TPM supporters tour the area, they are impressed with the changes. Sometime during the tour, however, one of the managers is likely to say, "This looks great, but we have spent a lot of money cleaning the equipment and training the employees in TPM principles and how to inspect their equipment. So, what impact is this making on our profit picture? After all, we can't stay in business with just clean equipment!"

Now panic sets in for the TPM coordinator. There is no clear an-

MAINTAINING THE TPM VISION

swer to this question. The soft benefits—better morale and better working conditions, for example—are now fading. Someone has asked the "show-me-the-money" question.

Unfortunately, in many companies, TPM has been suboptimized simply because no one had the vision to build a business case or a cost/benefit analysis when the program was first implemented. In fact, many companies begin TPM activities on the wrong equipment in their plants simply because the financial impact is not understood.

Early Focus

Where should initial TPM efforts in a plant focus? In order to maximize the return on investment that a company makes in TPM, it must focus on the critical equipment in the plant. Critical equipment is equipment that makes a significant different to the plant's operation if it operates and produces as it was originally designed. It might be an equipment item that is a production bottleneck or a piece of equipment that requires a high level of maintenance resources to keep running.

Once the critical equipment is identified, it is important to benchmark its performance with the OEE and project its financial role within the budget. Now, when the question arises about the impact TPM is having on profit, the answer is clear and in financial terms that managers will understand. Furthermore, if additional funding is required for additional tools, training, or personnel, the return on that investment is easy for anyone to calculate.

TPM is more than just another program that companies can implement. It is an operating philosophy that must be tied to the company's profit picture. TPM efforts that are not connected to the bottom line have little chance of success within a company. Although certain activities such as cleaning equipment and organizing the workplace are a part of TPM, long-term viability is doubtful unless TPM coordinators take a financial approach that highlights TPM's benefits.

Task Transfer to Operations

Certain tasks should not be transferred to operations personnel. Guidelines for these tasks are listed in Figure 15-1. Tasks that require special skills are usually beyond the scope of the training com-

> **TASKS BEYOND THE SCOPE OF AUTONOMOUS MAINTENANCE**
>
> - Tasks requiring special skills
> - Overhaul repair in which deterioration is not visible from the outside
> - Inspection and repairs that require significant disassembly
> - Tasks requiring special measuring techniques and tools
> - Tasks posing substantial safety risks
>
> **FIGURE 15-1**

ponent of a TPM program. For example, teaching operators to perform reverse dial indicator alignment on a coupling would probably be outside the scope of the TPM program. The focus should be on tasks that are basic, yet add value by relieving maintenance technicians to be redeployed on more complex tasks.

Another type of task that is beyond the scope of most TPM programs is one in which the deterioration is not obvious from a visual inspection. If it is not visual then special tools will be required to make the inspection, i.e., thermography or vibration analysis. Visual controls such as gauge marking can make most basic inspections easy for the operators. This is the type of task on which the majority of operator-based inspections should focus.

Inspections and repairs that require significant disassembly (such as major repairs and replacements) require skills and expertise beyond what most operators possess. Unless considerable training is to be provided the operators, they should not be involved in these types of activities. Realistically, they are still operators. If they undertake these activities, then who is going to operate the equipment? If the equipment is down long enough and often enough to warrant training the operators to this level of skills proficiency, then other problems exist within the company that need to be addressed.

> **OPERATIONS**
> **THE "NEW" JOB DESCRIPTION**
> - Deterioration Prevention
>   - Correct operation
>   - Clean, lubricate, tighten
>   - Make necessary adjustments
>   - Record breakdown and malfunction data
>   - "Team" with maintenance to work on improvements
>
> - Deterioration Measurement
>   - Perform daily inspections
>   - Work on larger periodic inspections
>
> - Equipment Restoration
>   - Perform minor repairs
>   - Report promptly and accurately on breakdowns
>   - Assist in breakdown repairs
>
> **FIGURE 15-2**

Tasks requiring special measuring techniques and tools can best be illustrated by the example above of the dial indicator. If an operator needs this type of specialized measuring tool to perform a task, then the tasks the operator are being expected to perform are too complex. You can not use a dial indicator once a year and still be proficient at reading and interpreting the results. This type of task should remain as the responsibility of the maintenance technicians.

Some tasks that pose substantial safety risks are not themselves inherently dangerous, but instead require special skills, tools, or knowledge to perform safely. For example, at one plant, production personnel took it upon themselves to reset and restart furnace fans that operated on 440VAC. Although pressing the reset did not seem dangerous in itself, an operator found the reset did not work. He opened the box and tried to force the starter to engage, then received severe electrical buns on his hands. Rather than innocently trying to fix something himself (shortening the downtime by not waiting for a maintenance technician), the operator was injured. Some tasks need to be clearly defined as off-limits to the operators for their own protection.

190 TOTAL PRODUCTIVE MAINTENANCE

Roles and Responsibilities

In a TPM work environment, some adjustments need to be made to the roles and responsibilities of both the operators and the maintenance technicians. Figures 15-2 and 15-3 highlight these new roles and responsibilities. These lists are not definitive in nature, but provide guidelines that may be utilized. For example, the first three items for the operators under deterioration prevention are basic PM tasks that they can be trained to perform.

Recording breakdown and malfunction data may be as simple as entering a code in a programmable logic controller (PLC) or distributed control system (DCS), logging an equipment malfunction. It may go as far as entering a work request for maintenance services in the company's CMMS/EAM system (see Chapter 9 for review). Keeping accurate equipment records is no longer the role of only the maintenance department. The shared ownership aspect of TPM requires that maintenance, operations, and engineering all be interested in keeping accurate equipment records.

MAINTENANCE
THE "NEW" JOB DESCRIPTION

- Improve Equipment Maintainability
    - Shorten repair times
    - Improve efficiency

- Train and Help Operators
    - Provide instruction
    - Prepare task sheets

- Advanced Functions
    - Research and apply maintenance
    - Technology
    - Set maintenance standards
    - Maintain accurate maitenance records
    - Evaluate maintenance activities
    - Provide input to engineering (design & spec.) Purchasing, operations, stores

FIGURE 15-3

MAINTAINING THE TPM VISION

As for the revised role of maintenance technicians, the activities listed in Figure 15-3 are already carried out by maintenance in companies, including non-TPM ones, where it has evolved to a mature status. This list of job roles shows a higher level of technical skills than typically found in those maintenance departments where fire fighting is the normal mode of operation.

Collecting data, processing it into useable information, and then providing technical feedback to engineering, inventory, purchasing, and operations becomes important. The equipment data collected and analyzed by the maintenance personnel provides critical information for companies trying to improve the OEE of mission-critical equipment. In this environment, the maintenance department truly becomes the doctor, not the paramedic.

Common Mistakes

Figures 15-4 and 15-5 highlight common mistakes encountered when implementing TPM. The first mistake in Figure 15-4 is the all-too-common reason encountered when companies are trying to reduce headcount. They view TPM as a way of eliminating maintenance technicians. They never see beyond task transfer from maintenance to operations. They do not have the technical understanding of their asset base. Thus, they do not realize that the maintenance department needs to ascend to another technical level. Instead, they view the freed up maintenance technicians as excess. Instead of re-

The Wrong Reason to Implement TPM Strategies

- Eliminating Maintenance Technicians
- To Give Control of the Maintenance Department to Operations or Production
- Just to have Maintenance Personnel on Teams

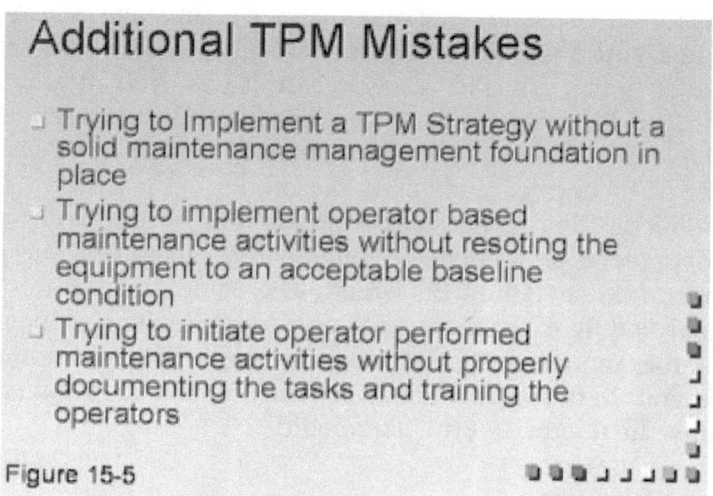

Figure 15-5

deploying them to fulfill higher-level technical responsibilities, they lay them off. Then in a few years when their equipment is encountering severe problems, they wonder what went wrong. Companies with a lack of technical expertise or a lack of respect for technical issues in management will almost always make this mistake.

The other two items in Figure 15-4 are relatively self-explanatory. Giving control of the maintenance department to operations or production is typically a knee-jerk reaction to an organizational problem. Disguising this under an attempt to implement TPM is an error which will doom any future effort to implement TPM. Both operations and maintenance will always refer to the past attempt and not cooperate with any future effort, whether for the right reason or not. Furthermore, simply putting maintenance personnel on teams and calling that process TPM will not succeed either. In plants where maintenance personnel have been put onto production teams without any technical focus, the maintenance personnel appear to be lost. They truly do not know what their job should be; therefore, their job devolves to a sort of handyman and they care only for mundane tasks that have no true technical focus.

Trying to implement TPM without a solid maintenance foundation (Figure 15-5) is typically a reaction to a situation where maintenance has not been effective and management has heard just enough about TPM to consider operator involvement in maintenance.

Management believes that having more bodies doing maintenance will make a difference. This approach never is successful. A solid PM program, inventory and purchasing practices, work order controls, etc., are all essential for gathering the data necessary to make TPM successful. TPM is an advanced manufacturing strategy, not a bandage.

Involving operators prematurely in TPM also leads to failure. If the equipment is not in an acceptable baseline condition, then the operators just become part of a larger fire department, but the fires are never really put out. Restoring equipment to an acceptable baseline condition is a critical prerequisite for beginning operator involvement.

Finally, the operators can not be empowered to take care of the basic maintenance of their equipment without clear instructions and detailed documentation. They also need to be trained to be proficient at performing the newly assigned tasks. The only one thing worse than no maintenance is poorly-performed maintenance. Unless the operators are given the skills necessary to perform maintenance correctly, the company will have an excessive amount of maintenance (operator) induced failures and the OEE will be negatively impacted.

TPM- The Only Reason

This chapter began with the question "If you could describe the main goal or objective for TPM in one statement, what would it be?". I hope that after reflecting on the material presented both in this chapter and throughout the entire book, your answer will match what is in Figure 15-6. If the OEE is the focus of a TPM effort, it will

TPM – The Only Reason

- To Increase or Improve the Overall Equipment Effectiveness
- "World Class" is perceived at 85%
- Where are YOU??

be a successful effort. If TPM is started for any other reason, it will be a failure. I hope that the material in this text will help you succeed.

INDEX

A
acceptance 181-182
activities 21-22, 59-78
adjustment losses 39-41
already-trained workers 146-147
analysis reports 131
apprentices 147
area organizations 65
attitude 31-44
autonomous groups 174-175
availability 45, 112-113

B
backlog 11-13, 66-67, 121, 126
breakdowns 31-39, 162-164

C
calculating costs 158-167
capacity 31-33, 126, 145-153
centralized organizations 64-65
checklists 94-95
cleaning 21, 33, 93-94
CMMS see computerized maintenance management system
combination organizations 65-66
communication 154
components 101-102
computerization 18
computerized maintenance management system (CMMS) 20, 129-144
condition based maintenance 99-100
continuing education 151
continuous improvement 24
control system 60-62
cost per repair 106
cost-benefit 4-5, 154-158
course outlines 151-153
craft technicians 66-67, 108, 122-124
critical spares 164-167
critical units 101
cycle count 114

D
data entry 138-9, 140-141
data transfer 141
decision tree 15-27
design cycle time 45
design life 37-38
downsizing 28
downtime 24, 48, 50

E
early equipment management 3, 23
effectiveness 16, 122-128, 177
efficiency 61-62, 122-128

energy cost reduction 86
equipment 2, 19, 30-49, 68-73, 83-84, 101, 119, 129
ERP-financial system suppliers 134
external setups 39-40

F
Failure and Effects Mode Analysis (FEMA) 23
financial indicators 177
financial optimization 23-24
function loss costs 155-156
future of TPM 181-183

G
goals and objectives 1-3, 59-60, 173

H
higher education 150-151
human error 38-39

I
idling 41-42
implementation 81-89, 138-143
information management 59, 64, 158
inspecting 21, 94
internal setups 40-41
inventory 13-14, 73-75, 85, 108-121

J
job plan 102-103
Just In Time initiatives 90

L
labor 84-85, 155
last issue date 113-114
life cycle costing 23
long-range plan 82-83
lost performance 16-162
lubrication 33-35, 93

M
maintaining TPM 184-194
maintenance 2-3, 6-14, 16, 33-35, 38-39, 59-68, 90, 112-121, 124-126, 133-134, 155-158, 167
maintenance information systems 129-144
major outage and overhaul savings 85-86
management techniques 172-174, 183
material costs 155
Mean Time Between Failure (MTBF) 24, 36
Mean Time To Repair (MTTR) 24, 36, 106
measuring equipment performance 44-46
measuring performance 82-83, 120-121
methodology 14-15
monitoring 171-172

N

nameplate information 70-72
normal equipment wear 36-37

O

obsolete parts 117
OEE see overall equipment effectiveness
oil analysis 22, 99
operational errors 38-39
operational standards 35-36
operator involvement 20-22
operator-based maintenance 169-172
organization 59, 64-66, 75-78
outages 96
overall equipment effectiveness 45-46, 47-58, 91, 106
overhead costs 155
overloads 41-42
overtime reduction 86

P

parts 73-75, 119-120
performance 45, 176-180
planning 18, 124-126
Predictive Maintenance (PDM) 22, 90-107, 156
preventive maintenance (PM) 13, 15, 90-107, 156, 159-164
purchasing 13-14, 86, 108-110, 115-118

Q

quality 43-44, 53, 86, 106
quantity 112-113

R

reactive work 16
rebuilds 96
reduced capacity losses 42-43
reliability centered maintenance (RCM) 23
reliability engineering 23, 37-38, 100
reliability of data 141
repair 156
replacements 24, 96, 156
resistance 99
restart losses 44
rewards 173-174
routines 4, 93

S

scheduling 18, 103-105, 125, 126-128
service level 120, 121
setup losses 39-41
shutdowns 96
software selection 134-139
staffing 59, 66-68
standardization 36-37, 115-116
startup losses 44
stockouts 120
stoppages 41-42
storeroom locations 116-117
system cost justification 83-87
system installation 138, 139-140

T

team-based maintenance 168-175
technical schools 149-150
technology 182-183
total economic maintenance 154-167
total productive maintenance 1-30, 57-58
total quality management 5, 90
training 3-4, 94, 137-8, 141-142, 146-153, 170-171
transferring tasks 169, 187-190
turnover 120

U

update records 138
upstream malfunctions 41-42
uptime 91

V

vendors 133-139, 150
vibration analysis 22, 98
vocational schools 147, 149-150

W

warranty costs 86
work orders 17-19, 21, 59, 62-63, 95, 105, 132
world class inventory 111-112

Z

zero breakdowns 33-39

www.ingramcontent.com/pod-product-compliance
Lightning Source LLC
Chambersburg PA
CBHW021125300426
44113CB00006B/294